AF840693

Sustainable Innovations and Performance of Organizations in the Age of Digitalization

The digital era leverages blockchain and IoT technologies to revolutionize sustainability by enabling real-time resource monitoring, reducing waste, and providing transparent supply chain verification. This book examines sustainability and organizational transformation through comprehensive analysis supported by tables, charts, and references, offering practical guidance for modern sustainable business practices.

Sustainable Innovations and Performance of Organizations in the Age of Digitalization presents how blockchain technology enhances transparency and traceability in supply chains, ultimately verifying the sustainability of sourcing practices and materials. Additionally, it discusses how digital platforms contribute to the development of circular economy models by enhancing the tracking and recycling of products throughout their life cycle. The book also highlights innovative solutions that enable businesses to adopt more sustainable practices, thereby reducing their environmental impact. By examining sustainability and organizational challenges, this book offers to bridge the gap between theory and practice.

A valuable resource for students, graduates, educators, researchers, and professionals in logistics, supply chain management, green logistics, production, leadership, organization, sustainability, and digitalization.

Industrial and Systems Engineering Series

Series Editosr: Waldemar Karwowski, University of Central Florida, Orlando, USA & Hamid Parsaei, Texas A&M University

Industrial Engineering has evolved as a major engineering and management discipline, the effective utilization of which has contributed to our increased standard of living through increased productivity, quality of work, and quality of services and improvements in the working environments. The Industrial and Systems Engineering book series provides timely and useful methodologies for achieving increased productivity and quality, competitiveness, globalization of business and for increasing the quality of working life in manufacturing and service industries. This book series should be of value to all industrial engineers and managers, whether they are in profit motivated operations or in other nonprofit fields of activity.

Managing Professional Service Delivery: 9 Rules for Success
Barry M. Mundt, Francis J. Smith, and Stephen D. Egan Jr.

Mobile Electronic Commerce: Foundations, Development, and Applications
Edited by June Wei

Manufacturing Productivity in China
Edited by Li Zheng, Simin Huang, and Zhihai Zhang

Human Factors in Transportation: Social and Technological Evolution Across Maritime, Road, Rail, and Aviation Domains
Giuseppe Di Bucchianico, Andrea Vallicelli, Neville A. Stanton, and Steven J. Landry

Laser and Photonic Systems: Design and Integration
Edited by Shimon Y. Nof, Andrew M. Weiner, and Gary J. Cheng

Supply Chain Management and Logistics: Innovative Strategies and Practical Solutions
Edited by Zhe Liang, Wanpracha Art Chaovalitwongse, and Leyuan Shi

Production Management, Manufacturing, and Process Control
Edited by Beata Mrugalska, Waldemar Karwowski, and Tareq Z. Ahram

Industry 4.0 and Intelligent Manufacturing Technologies
Edited by Beata Mrugalska, Waldemar Karwowski, and Tareq Z. Ahram

Sustainable Innovations and Performance of Organizations in the Age of Digitalization
Edited by Sonia Umair, Umair Waqas and Beata Mrugalska

For more information about this series, please visit: www.routledge.com/Industrial-and-Systems-Engineering-Series/book-series/CRCINDSYSENG

Sustainable Innovations and Performance of Organizations in the Age of Digitalization

Edited by
Sonia Umair, Umair Waqas and
Beata Mrugalska

Routledge
Taylor & Francis Group
NEW YORK AND LONDON

First published 2026
by Routledge
605 Third Avenue, New York, NY 10158

and by Routledge
4 Park Square, Milton Park, Abingdon, Oxon, OX14 4RN

Routledge is an imprint of the Taylor & Francis Group, an informa business

© 2026 selection and editorial matter, Sonia Umair, Umair Waqas and Beata Mrugalska; individual chapters, the contributors

The right of Sonia Umair, Umair Waqas and Beata Mrugalska to be identified as the authors of the editorial material, and of the authors for their individual chapters, has been asserted in accordance with sections 77 and 78 of the Copyright, Designs and Patents Act 1988.

All rights reserved. No part of this book may be reprinted or reproduced or utilised in any form or by any electronic, mechanical, or other means, now known or hereafter invented, including photocopying and recording, or in any information storage or retrieval system, without permission in writing from the publishers.

For Product Safety Concerns and Information please contact our EU representative GPSR@taylorandfrancis.com. Taylor & Francis Verlag GmbH, Kaufingerstraße 24, 80331 München, Germany.

Trademark notice: Product or corporate names may be trademarks or registered trademarks, and are used only for identification and explanation without intent to infringe.

ISBN: 978-1-041-00841-5 (hbk)
ISBN: 978-1-041-00842-2 (pbk)
ISBN: 978-1-003-61190-5 (ebk)

DOI: 10.1201/9781003611905

Typeset in Times
by KnowledgeWorks Global Ltd.

Contents

PART VI *Education Institutions and Sustainability*

Preface

In the digital age, sustainable innovations and performance are greatly impacted by blockchain technology and the Internet of Things (IoT). IoT allows for real-time monitoring and management of resources like water and energy, improving efficiency and reducing waste. Smart grids, for example, can optimize electricity usage, leading to lower emissions and cost savings. Blockchain technology ensures transparency and traceability in supply chains, which helps to verify the sustainability of sourcing practices and materials. Digital platforms also help to develop circular economy models by improving the tracking and recycling of products throughout their life cycle. These innovations can help businesses adopt more sustainable practices by lowering their environmental impact.

This book will be of interest to students, graduates, teachers, researchers, and general workers in logistics, supply chain, green logistics, green supply chain production, management, leadership, organization, sustainability, and digitalization. It will help to support courses on sustainable performance of organizations for graduate programs including business, management, supply chain management, and industrial engineering.

The 15 chapters allow to have a deep look at sustainability and organizational problems and make a bridge between theory and practice. Each of the chapters has its own place and is supported with tables, charts, and illustrations, and a wide list of bibliographic references.

About the Editors

Dr. Sonia Umair is Assistant Professor at Dhofar University, Oman, and a distinguished researcher in the field of management. She is a Chartered Management and Business Educator (CMBE) and Fellow of the Higher Education Academy (FHEA–UK). With over a decade of experience, she has conducted extensive research in HRM, CSR, organizational behavior, and sustainability. Dr. Sonia has published widely in reputable journals and actively engages in scholarly projects that address contemporary challenges in management and sustainable business practices. Her work emphasizes evidence-based insights, rigorous methodology, and the integration of theory with practical relevance. Through her research, she contributes to advancing knowledge, developing innovative solutions to organizational challenges, and inspiring both academics and practitioners to adopt effective, sustainable, and ethical management practices globally.

Dr. Umair Waqas is Assistant Professor at Dhofar University, Oman, with over ten years of experience in research and education. He is a Chartered Management and Business Educator (CMBE) and Fellow of the Higher Education Academy (FHEA–UK), his research focuses on supply chain management, green digital servitization, circular economy, sustainability, and resilience engineering. He has authored several book chapters in *Sustainable Innovations and Performance of Organizations in the Age of Digitalization* (Taylor & Francis) and contributes to the *Oman Observer* on topics related to Oman Vision 2040, sustainable logistics, and tourism development.

Dr. Beata Mrugalska, Ph.D., D.Sc., Eur.Erg., is Associate Professor, Vice-Dean for Science (2024–2028) and Head of the Division of Applied Ergonomics (2020–2024; 2024–2028), Institute of Safety and Quality Engineering, Faculty of Management Engineering, Poznan University of Technology in Poland. Since 2018, she has been a board member of the Center for Registration of European Ergonomists (CREE). She is also a board member of the Polish Academy of Arts and Sciences, Commission of Ergonomics, and a member of the Ergonomics Committee at the Polish Academy of Sciences, Branch Poznan. Dr. Mrugalska is the author or co-author/editor or co-editor of 15 books and more than 150 journal articles, book chapters, and conference papers. She serves as a member of the journal editorial board of several journals and is a co-editor of the Taylor & Francis series "Body of Knowledge in Human Factors and Ergonomics."

Contributors

Zaheer Ahmed Khan
Mazoon College
Muscat, Oman

Faiza Amir
Mazoon College
Muscat, Oman

Ihsan Ashiq
Mazoon College
Muscat, Oman

Zeeshan Asim
College of Business, University of Buraimi
Al Buraimi, Oman

Ujjal Bhuyan
Department of Commerce, Jagannath Barooah University
Jorhat, India

Ilya Bystrov
Faculty of Business and Management, American University of Phnom Penh
Phnom Penh, Cambodia

Brigita Gajšek
University of Maribor
Maribor, Slovenia

Muhammad Zulkifl Hasan
Faculty of Information Technology, University of Central Punjab
Punjab, Pakistan

Mohamed Wahish Hashim
College of Business, University of Buraimi
Al Buraimi, Oman

Asghar Hayat
Department of Business Administration, Ghazi University
Dera Ghazi Khan, Pakistan

Muhammad Zunnurain Hussain
Department of Computer Science, Bahria University Lahore Campus
Lahore, Pakistan

Shahid Iqbal
Bahria University
Islamabad, Pakistan

Shoaib Irshad
Department of Business Studies, Namal University Mianwali
Punjab, Pakistan

Shad Ahmad Khan
Faculty of Business and Management American University of Phnom Penh Phnom Penh, Cambodia
Phnom Penh, Cambodia

Shahzad Ahmad Khan
College of Business, University of Buraimi
Al Buraimi, Oman

Muzammil Khurshid
University of Punjab, Gujranwala Campus
Punjab, Pakistan

Beata Mrugalska
Poznan University of Technology
Poznan, Poland

Devarajanayaka Muniyanayaka
College of Business, University of Buraimi
Al Buraimi, Oman

Zia Ur Rehman
Department of Business Administration, Ghazi University
Dera Ghazi Khan, Pakistan

Karima Sayari
Department of Managerial and Financial Sciences, Al Zahra College for Women
Muscat, Oman

Ibrahim Rashi Al Shamsi
College of Business, University of Buraimi
Al Buraimi, Oman

Adnan Ahmed Sheikh
College of Business, University of Buraimi
Al Buraimi, Oman

Yogesh Sinha
Ohio University
Ohio, USA

Veeraiya Perumal Sriram
College of Business, University of Buraimi
Al Buraimi, Oman

Simona Šinko
University of Maribor
Maribor, Slovenia

Sonia Umair
College of Commerce and Business Administration, Dhofar University
Dhofar, Oman

Umair Waqas
College of Commerce and Business Administration, Dhofar University
Dhofar, Oman

Muhammad Ziaullah
Department of Business Administration, Ghazi University
Dera Ghazi Khan, Pakistan

Part I

Foundational Concepts

1 Future Trends in Sustainable Digital Innovation

Muhammad Zulkifl Hasan, Muhammad Zunnurain Hussain, Umair Waqas, and Sonia Umair

1.1 INTRODUCTION

1.1.1 The Shift towards Sustainable Digital Innovation

The modern economic growth depends heavily on digital innovation because it remoulds industrial infrastructure alongside service systems and social frameworks. Fast technological progress creates enhancements in energy consumption as well as environmental resource destruction and digital inequality problems. The world has chosen sustainable digital innovation as its base principle for digital technology transformation where all stages from creation to distribution to administration serve to protect natural health while securing satisfaction from society and sustainable economic stability (Omol, 2024). The adaptation represents an active approach rather than a passive method of response. People acknowledge the urgent necessity of innovation to address technical and market issues, ethical and ecological sustainability requirements, and global sustainability needs (Trevisan et al., 2023). Technology development needs total implementation of sustainability principles from initial idea generation to product creation and later user phases until reaching end-of-life. Digital progress is entering a sustainable path through joint implementation of sustainable development goals by governments and institutions with corporate partners (Afeltra et al., 2023; Yin et al., 2023).

1.1.2 Overview of Emerging Technologies and Sustainable Goals

The innovative tendencies in the digital technologies sphere are providing the unique chances to work on the progress of the goals of sustainable development more actively. It is important to state that artificial intelligence (AI), the Internet of Things (IoT), blockchain, edge computing, and quantum technologies are being discussed not only as opportunities for increasing efficiency in industries or boosting economic growth but also as the drivers of the SDGs. Due to these technologies, efficiency of monitoring the environmental systems, delivery of services, application of resources, and transparency in governance are enhanced. For instance, AI can predict potential risk such as environmental threats; IoT ensures real-time monitoring of resource usage at

DOI: 10.1201/9781003611905-2

various strata; and blockchain brings about tamper-proof records for carbon credits or ethical sourcing. When used in support of sustainability, these technologies can redefine the solutions that societies need to mobilize in order to approach climate change, water and food management, shift to renewable energy, and create sustainable infrastructures. The fusion of individuals, augmented intelligence/machine learning, and sustainable goals are not only a fusion of technologies but also of cultures driven by collaboration, ethical standards, and policies that adopt the tenets of sustainability within the advancement of technology (Abad-Segura et al., 2020).

1.2 NEXT-GENERATION DIGITAL TECHNOLOGIES AND SUSTAINABILITY

1.2.1 Artificial Intelligence and Sustainable Practices

The emerging revolutionary power of AI serves to reach sustainability goals. Intelligent systems based on AI analyse significant amounts of data to detect recurring patterns for prediction techniques that serve industries managing energy power as well as farming activities, urban development, and supply chain operations (Hrustek, 2020). Utilities achieve higher power grid efficiency by using AI models to predict energy consumptions and manage power systems, which reduces power waste and optimizes exploitation of renewable resources. Through AI-controlled precision farming operations, farmers conduct environmental data analysis to automate their irrigation and fertilization systems, thus achieving better crop output and reduced chemical usage. Urban areas achieve benefits from AI controllers for waste management and traffic control, which reduce both traffic congestion issues and waste resources. AI training operations produce sustainability issues because they require extensive amounts of energy during the process. Massive model training for AI produces significant carbon emissions that drive researchers to establish "Green AI" methods to develop sustainable hardware systems and algorithms which require less energy use. AI integration into sustainability initiatives needs urgent study to solve environment-related side effects since these positive societal outcomes possess related implications (Ghobakhloo et al., 2021; Satalkina and Steiner, 2020).

1.2.2 The Evolution of IoT for Environmental Monitoring

IoT technology transforms information collection about the environment and routes them through processing before executing tasks via its networked devices. The IoT creates global networking between trillions of devices that include forest and river sensors together with residential smart meters to offer ongoing real-time environmental sensor data. The performance of IoT monitoring devices helps detect air and water conditions along with noise pollution, temperature, and humidity to create more specific protective approaches for environmental spaces. The integration of IoT technology enables urban areas to develop adaptive lighting systems, traffic management systems, and waste management solutions that lower carbon emission levels. Water conservation in agricultural fields enables soil sensors to work with connected irrigation systems, which results in high crop productivity for farmers. Many crucial security risks appear due to extensive IoT system deployment. The combination of production methods and disposal practices for billions of devices produces environmental risks from e-waste contamination and continuous data transfers result in energy consumption problems and data security. The support of sustainability needs complete IoT deployment to develop standards that establish device lifecycle management, data security practices, and interoperability rules.

1.2.3 Blockchain's Potential in Carbon Tracking and Offsetting

Businesses find blockchain technology effective for managing sustainability goals because it helps them track carbon footprints more openly. Decentralized ledgers in blockchain create permanent copies of carbon emission data that prove environmental improvements while stopping any repeated use of data or dishonest activities. This system helps us meet official requirements and make sure that benchmarks are reliable for carbon market buyers. Carbon market platforms that sell digital tokens with credits let more people take part in sustainability investment opportunities. The combination of blockchain and IoT forms a monitoring system that simultaneously measures emissions and starts emission offset programmes without delay. Some consensus methods in blockchain need a lot of power and face questions about their impact on sustainability. Modern proof-of-stake technology helps solve blockchain energy issues and its wider adoption will make blockchain work towards environmental targets. The blockchain technology creates reliable platforms for building systems that manage the environment fairly and easily at a large scale.

1.3 GREEN COMPUTING AND DATA CENTRES

1.3.1 Energy-Efficient Data Centres

But data centres are being used to keep going in the digital age and along the way they eat up much energy. The data centres increasingly consume energy, but more cloud computing platforms and data services are asking us to find effective solutions to use less energy in data centres. Modern-day green data centres use solar and wind energy systems to power this century's modern heat-dissipation technologies (such as liquid immersion and free air technologies systems). Virtualization and the agile resource controls combine with AI-based workload management to reduce wasted energy when servers are using power but are not processing tasks – smart energy functions. Cool regional climates are the place where the data centres would be chosen, as they require less energy to be cool. Data centre efficiency assessment has become an essential measure of the instrument to calculate the power usage effectiveness. Several providers are planning to create data centres that use less carbon dioxide than the emission from all of their servers, and major cloud companies are set to make their data centres neutral in terms of carbon emissions. To run these highly effective data centres, however, they will also have to remain within the larger strategy of expanding their respective capacity to accommodating future data growth through edge computing systems, renewable power connections, as well as smart power consumption techniques.

1.3.2 Advances in Green Computing Technologies

The conceptualization of green computing is the strategic change towards designing computing systems which uses less energy, produce less heat, and minimizing the use of energy from the cradle to the grave. Innovations in green computing span both hardware and software. Energy-efficient processors, biodegradable circuit boards, and modular architecture support easy repair and upgrade on the hardware side. This reduces the material that has been consumed and prolongs the life of devices. Specifically, we develop green algorithms for computing with low energy use at both the software level and in the cloud services, for instance, allowing users to choose energy-oriented virtual machines based on cloud services. Other new emerging technologies such as neuromorphic computing and quantum computing may be possible to reduce energy needs at brain-like efficiency or utilize quantum phenomenon to reduce energy needs. Green computing has turned out to be a necessity rather than an environmental imperative, owing to business necessity: escalating energy costs, regulatory concerns, and the growing demand for sustainable technology products by consumers.

1.3.3 Edge Computing and Its Role in Reducing Carbon Footprint

Edge computing makes data processing better by placing processing resources close to data origins to cut back on cloud server network use. The upgraded network system produces environmental benefits to everyone. Edge computing handles data near its source on either local machines or nearby systems which prevents power consumption on distant data centre connections. Edge computing processes materials now to optimize smart city technology which reduces carbon emissions from moving vehicles and industrial plants. These important sustainability features deliver their services without delays when they need to work during emergencies. The power requirements of edge devices remain modest when they lower the energy demands of traditional centralized systems especially with energy-saving designs. Sustainable design requirements for edge computing devices need to cover every part of their life cycle to achieve maximum carbon savings across the entire technology field.

1.4 THE RISE OF CIRCULAR ECONOMY IN TECHNOLOGY

1.4.1 Recycling and Reuse of Technological Components

The digital technology sector now follows new concepts instead of continuing with its existing take-make-dispose system. Under a circular model resources stay in use by making products strong enough to reuse and recycle time after time. To improve technological components, we need to change how electronic devices are made delivered to users then properly ended. Changing technology and customer preferences for new products create serious problems by increasing electronic waste (Salvi et al., 2022). Every year millions of smartphones, laptop, and server devices end up as waste while still holding meaningful materials like gold, copper, lithium, and rare earth elements. Companies now use modern robotics and AI tools to upgrade their recycling systems while replacing hazardous extraction processes. Organizations are spending money on facilities that fix used products back to original standards and keep them operational longer while lowering the requirement for obtaining natural resources. Manufacturers need to create products that stay editable and lasting through policymakers who make business owners repair products responsibly (Liu et al., 2023).

1.4.2 Digital Innovations in Waste Management and Resource Recovery

Our digital systems help us run waste management better while creating sustainable outcomes. Sensors in modern cities monitor dumpster waste levels to help government services create better delivery routes that prevent environmental damage from unnecessary driving. The AI technology system makes recycling plants operate better by handling materials properly. AI technology improves waste quality standards while turning more waste into marketable products. The IoT gives industrial companies instant access to waste operations data to locate material loss patterns and develop methods to recycle unused materials. By using network platforms businesses develop industrial connections that turn industrial waste into useful materials. Business and community users sell and purchase unused items through Internet-based exchanges. A safe monitoring system tracks every item through its path from producer to administrator. Our updated waste management technology helps the present generation stop creating waste by returning materials to use.

1.4.3 Sustainable Product Lifecycle Management

Sustainable PLM systems are required for organizations to incorporate environmental factors at every step from product life beginning to manufacturing then to customer use and ending with end-of-life processing. Some special PLM digital tool is used to help the organization study and track product performance to enable product selection to minimize environmental impacts. Replicas of the product's digital twins allow testing of performance relieving the products prior to failure and monitoring usage for lifespan planning to save resources. The actual measurements of product health are provided by devices which leave sensors on the product and push a measurement to a user to decide on the control of a product function, before something fails. Through carbon footprint, software platforms highlight leaders that convey how things that leaders choose to use – like designing and material selection – lead to increased or reduced carbon emissions in terms of energy use. The systems leverage AI technology to suggest eco-pleasant design solutions and identify issues with the product's production and delivery. Reduced costs and a smaller environmental impact, alongside increased supply chain security and the best attempt to meet customers' wants for eco-friendly products, are all as a result of company efforts to successfully manage products sustainably.

1.5 AI AND MACHINE LEARNING FOR PREDICTIVE SUSTAINABILITY

1.5.1 Using AI to Forecast Environmental Impact

For forecasting and mitigating environmental impact, synthetic intelligence is vital as it can analyse very huge, unclear information which typical fashions cannot interpret at equal rate or accuracy. AI systems that build on satellite imagery, climate models, sensor networks, past environmental data, and even on site that can predict floods, wildfires, droughts, pollution spikes, have long been created. If allowed, governments, industries, and communities can respond in a more informed and proactive way to protect against harm and use their resources more effectively by preparing and planning. These predictions give us first level insight. By improving on climate modeming's ability to use equations to rapidly simulate the future under alternative therapies and behaviours, equations leaders can make more well-informed decisions. We are moving into the original environmental interest assessment in living and observing assessments of projects and their ecological figure changing from labour tense and quick to dynamic. However, it is not the analytical capability that AI has in its set of advantages in this domain, but rather, it is the ability to adapt to new data, and to evolve as the new environmental dynamics change. However, these models are free of bias, transparent, and inclusive but remain to be legitimate, and effective.

1.5.2 Machine Learning Models for Energy Optimization

Machine learning algorithms are learning usage patterns, predicting demand, and optimizing resource allocation, driving energy system towards a new efficiency. For an example of such will be taking the case of buildings, where machine learning is making the settings of the systems tuned up to take the number of people there and whether the weather is expected to be cold or hot, for heating and cooling respectively. Smart meters and grid analytics use machine learning to detect anomalies in machine consumption for energy loss reduction as well as suggest behaviours for the user. For instance, machine learning is used to enhance the placement of wind turbines on top of solar panels and battery storage in wind turbines to maximize output and grid stability. Using machine learning in an industrial scale across the whole manufacturing process the aim is to find inefficiencies, make suggestions to help overcome these inefficiencies and to support lean manufacturing practices with lower c. These models are themselves learned and self-sufficient, created to update parameters with new data arriving on them and operating autonomously to become intelligent energy ecosystems that are adaptive and sustainable. Due to these technologies becoming more tractable and user friendly they are opening up opportunities for even small-scale users to select intelligent optimization solutions that will reduce their carbon footprint.

1.5.3 Autonomous Systems for Efficient Resource Use

Autonomous systems, including drones, robots, or machines controlled by AI, are taking over the management of resources in different spaces with the goal of saving

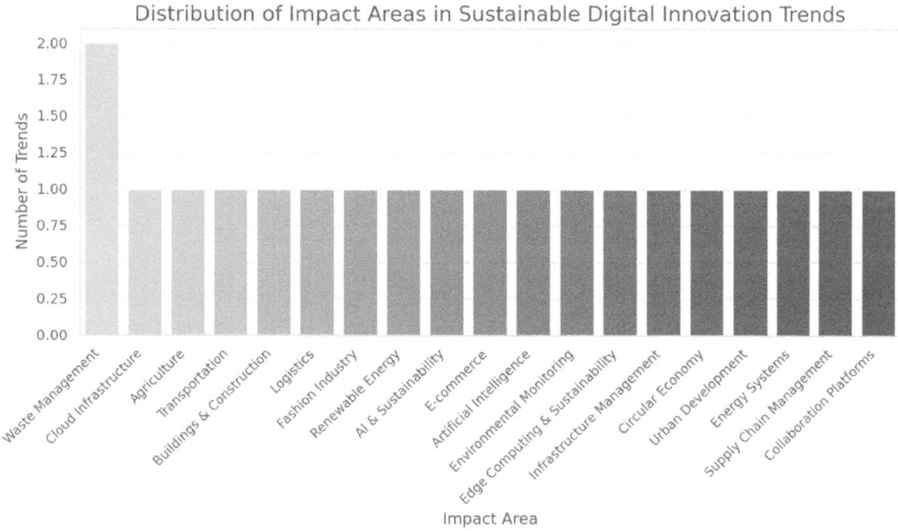

FIGURE 1.1 Impact area in sustainable digital innovation. (Lastman, 2025b)

resources and conserving. Autonomous tractors and irrigation grind conveyors can utilize data from soil sensor and forecasts on projected weather to rapidly cut waste and runoff by giving just the appropriate sum of water and nutrients. In logistics, autonomous delivery vehicles and drones offer a real time compensation of routes to minimize fuel consumption and emissions. Industrial operations using autonomous robots require less material waste and downtime producing a more precise and fewer error jobs as in Figure 1.1. They are running continuously and adaptive and do usually better than humans in situations where efficiency and consistency are the main objectives. Furthermore, integration of the devices in the cloud data system enables coordination of the whole system, for example, for group of autonomous bot, the resources are under control of one hub for energy saving. At the same time, the lifecycle, from production, consumption of energy, and disposal, must be evaluated to ascertain whether autonomous systems in fact in fact enable long-term resource efficiency (Hu et al., 2023).

The above graph depicts the frequency rate of impact areas within the sustainable digital innovation trends to allow taste of trends by sectors. On the y-axis there is a count of how many trends were identified, and on the x-axis there are categorizations of the impact area, which are waste management, cloud infrastructure, agriculture, and so on. Waste management is the most appeared impact area as a trend while cloud infrastructure occupies the second place; however, all categories differ by the level of trend appearance. For instance, sectors like renewable energy, AI, e-commerce, and others are almost equally represented while there are some sectors that seem to be represented few times, such as collaboration platforms and supply chain management. It is evident from the graph that sustainable digital innovation trends are vast and differ in the degree of implementation in various impact fields.

The above graph shows the list of major challenges that have been correlate with sustainable digital innovation and the occurrence of each challenge. On the y-axis

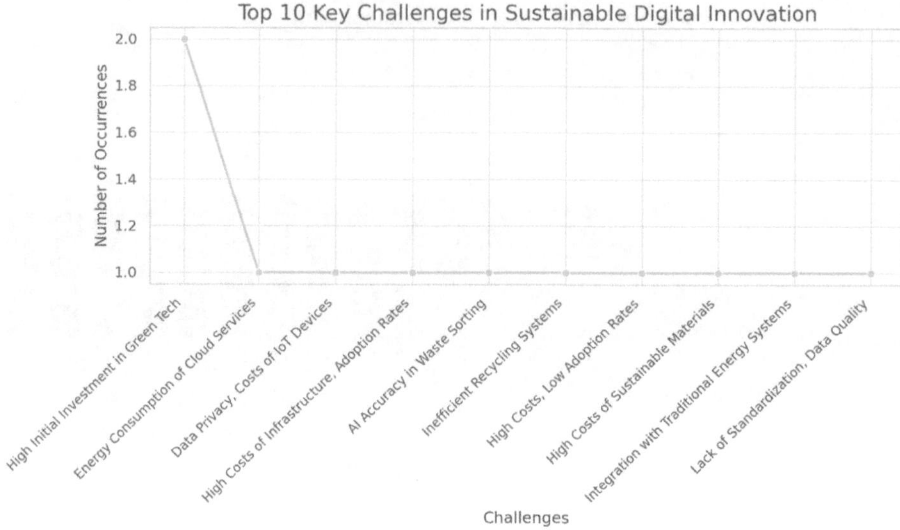

FIGURE 1.2 Top 10 key challenges. (Lastman, 2025b)

there are presented the number of occurrences while on the x-axis there are presented several challenges. From the analysis of the numbers shown above, the most common challenge is "High Initial Investment in Green Tech", inclusive of two. The remaining issues, namely, "Energy Consumption of Cloud Services", "Data Privacy", and "High Costs of IoT Devices" are less frequent with an appearance of only once and are thus deemed to be important in the field of sustainable digital innovation as indicated in Figure 1.2. This distribution shows that initial investments in green technology are considered as the biggest hurdle out of all the challenges.

The graph below shows the specification of the top ten technologies attached to sustainable digital innovation and their frequency. On the y-axis the bars indicate the number of occurrences, and on the x-axis, it gives the technologies as shown in Figure 1.3. Among those, AI is the most frequently mentioned technology, which is closely followed by IoT and big data in terms of the number of occurrences. Other technologies that also feature a high number of mentions are listed in the following order: these include sensors, blockchain, and renewable energy In addition, smart sensors, digital platforms, eco-friendly packaging, and battery storage also made appearance though they had fewer occurrences compared to the other technologies. This distribution regards the AI technology as the most significant global driver of sustainable digital transformation along with IoT and big data, while other technologies also define the emerging technologies mentioned here.

The graph represented how different trends are distributed in relation to their sustainability focus areas and how often each of them appears. The vertical axis defines the number of trends, while the horizontal axis is used to classify the focus areas of sustainability. Some of the most commonly occurring keywords include "Cutting Down Carbon Emissions in Data Centres", followed by "Efficiency in power use"

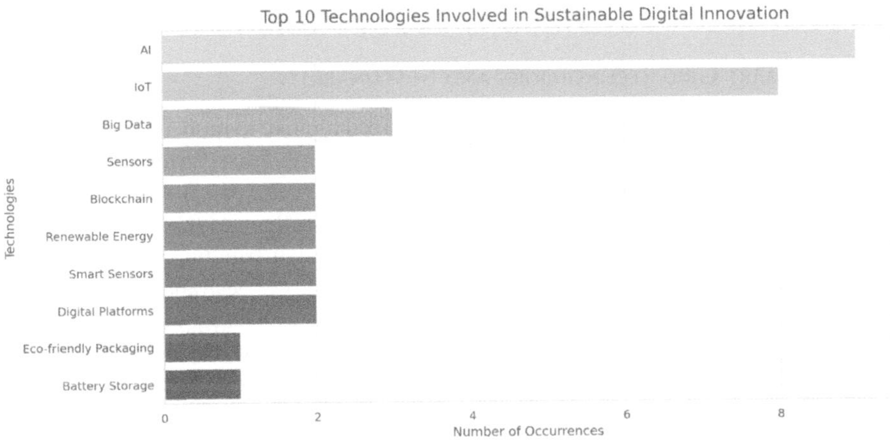

FIGURE 1.3 Top 10 technologies. (Lastman, 2025b)

and "Minimizing Waste" most of the time. Other critical dimensions of sustainable solutions with a high level of representation include energy efficiency in commercial buildings as well as reducing waste through intelligent sorting and recycling. On the other hand, the terms "Enabling Collaboration for Green Projects" and "Circular Product Life Cycle" are less common, indicating that sustainability is now a concern for innovation trends, but it is focused on different aspects in trends with less emphasis on creative progress as seen Figure 1.4.

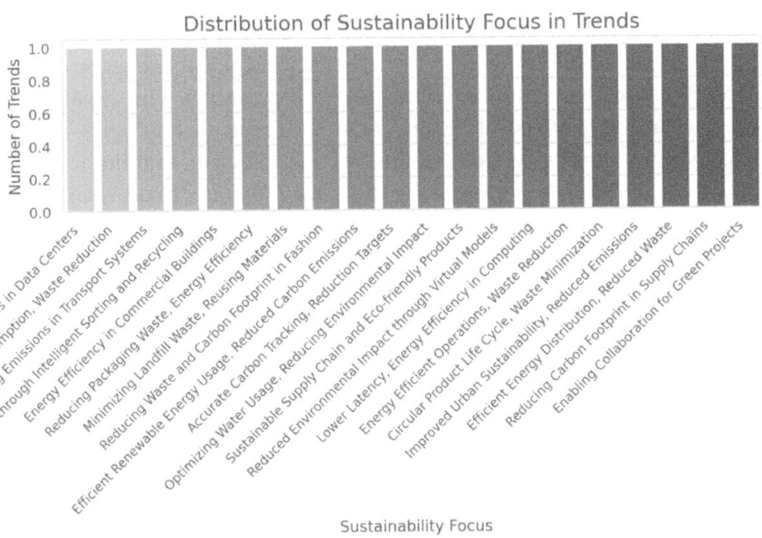

FIGURE 1.4 Distribution of sustainability in trends. (Lastman, 2025b)

1.6 FUTURE OF RENEWABLE ENERGY INTEGRATION

1.6.1 Smart Grid Technologies and Sustainability

The fundamental change in how energy is managed relative to more adaptively, more resilient or more sustainable, is smart grid technology. Smart grids employ digital communication, sensors, automation, and analytics to keep track of electricity flows, match supply and demand, and react to any disturbances (Geng et al., 2023). Compared to the one directional energy and information distribution mode, the smart grids enable two directions of energy and information flow. Desirable for integrating decentralized and intermittent renewable energy sources: solar and wind. A smart grid can plan for generation, dispatch the energy storage and dispatch the schedule of the power for a renewable power to be used reliably and efficiently (Kabeyi and Olanrewaju, 2023). Consumers are also prosumers of both producing and consuming electricity as part given to demand response programme where a consumer can be reimbursed for changing their usage to the off [peak time] period. However, in addition to that, improving the energy efficiency also results in overall system emissions. Smart grids are also supportive to electrification of other transport and heating sectors other than electricity and thus, these other sectors can be more decarbonized through digital and energy management (Hassan et al., 2024).

1.6.2 Digital Solutions for Managing Renewable Energy Resources

Modern computers enhance the performance of renewable energy systems by using AI. Our tools show detailed information about turbine and solar panel operations plus their internal components for immediate performance evaluation by operators. Actual conditions reported by sensors help identify when equipment parts break down and keeps them working longer. The integration of weather data with machine learning empowers operators to apply battery storage best when it produces most power while maintaining power consistency. Renewable energy developers use GIS tools because they combine topographic information with weather conditions plus access data to plan power plant creation. These technologies increase renewable energy performance while lowering production costs making it produce electricity less than burning fossil fuels. New digital networks need security and protection systems against cyberattacks that target the energy industry.

1.6.3 Blockchain's Role in Energy Trading and Management

Blockchain technology moves towards decentralized systems by transforming energy trading and control processes. Blockchain technology helps customers trade electrical energy directly from their solar power and wind power systems. Spontaneous smart contracts handle transactions from start to finish without managers and help users reduce payment fees and see what occurs. The blockchain network allows renewable energy providers to build unalterable evidence of energy source origins, which helps standard carbon trading marketplaces and corporate sustainability disclosures operate better. The blockchain system controls microgrids and rewards user power usage by logging energy usage and emission information transparently for all grid participants. The sustainability of blockchain systems relies first on how much energy their basic design requires. The blockchain needs energy-efficient consensus methods to show real value in building sustainable energy solutions.

1.7 CHALLENGES AND ETHICAL CONSIDERATIONS IN SUSTAINABLE INNOVATION

1.7.1 Ethical Implications of AI in Sustainability

AI has the power to make sustainability better, but the ethical standards must apply to all people to make proper use. This main issue, however, is that the algorithms built tend to be biased and can yield discriminatory results in environmental applications. Predictive police technology which combats crime and manages energy also contributes to the pressures placed on low-income neighbourhoods, by discouraging their social problems to continue. When the systems rely on biased training data sets, they may choose to supply resources to them first to more affluent zones. However, many AI systems display black box behaviour, which means people cannot see how these systems work and thus introduces problems of transparency and responsibility. The problem with AI systems is that when they are being used in such manners to aid in official decisions, public acceptance is contingent upon high transparency. Ethical AI systems ensure that the procedures are clear, where fairness is supported as something that is easily understandable, practices are inclusive and human support is needed to prevent harm. Control over AI has to be based on advanced technical knowledge and a strong ethical commitment to fairness and inclusionary standards as AI is used for sustainable purposes.

1.7.2 Addressing Data Privacy and Security in IoT-Driven Sustainability

A key ethical issue regarding IoT technology-based sustainability implementation emerges during its adoption stage between maximizing data usefulness and protecting privacy. Newly developed sensor networks throughout homes and urban areas and transportation systems and agricultural operations lead to the creation of large volumes of useful data that help optimize energy consumption and environmental tracking and public service delivery. The gathered data includes specific personal information about what individuals do and where they go together with their preferences. Topic-specific insight derived from combined IoT data analysis creates potential threats because authorities or corporations could achieve unauthorized surveillance capabilities and data vulnerabilities when proper security measures are absent. The exploitation of smart meters allows potential identification of personal routines through the analysis of energy use in homes. The ethical use of IoT for sustainable operations requires encryption and strong protection mechanisms for anonymized data which should accompany policies that detail the entire data collection process. The GDPR in Europe ban be viewed as examples of advancement in privacy protection yet they need modernization to maintain relevance in today's evolving smart interdependent network. The continuing existence of IoT sustainability initiatives combined with their long-term sustainability relies on maintaining strong privacy standards and security protocols as well as user autonomy protections.

1.7.3 Managing Digital Waste and E-Waste Recycling

The electronic innovations become a key to sustainability while concurrently posing as a major environmental crisis due to the emergence of electronics waste stemming from large-scale disposal of electronic devices. Yet consumer spending on new electronic products means short product lives and so trigger consumers to upgrade electronic waste at a million of tons per year as electronic waste originates from outdated devices. Among other things, deep inside electronic devices, it is lead, mercury, and even cadmium that end up turning into a serious threat to health and nature if we send those devices to the wrong disposal facility. First, these are materials that contain both harmful substances detrimental to human life and the Earth, in addition to rare earth elements and metal resources that can be reused. The biggest challenge regarding this is the creation of a sustainable e-waste management system through design principles as well as infrastructure development and public education. Manufacturers must develop products which can easily disassemble, maintain, and return to raw materials owing to the empirical design principles. Law focused on ending producer responsibility at a regulatory level force companies to practice responsibility in all stages for the life of its products, including the end-of-life stage. The dual purpose that digital platforms cater to is to optimize waste collection and be able to help businesses and consumers locate certifiable services to get rid of e-waste. Most of the global e-waste exports go to developing nations through which informal recycling activities are causing extreme environmental destruction and harm to humans in their territories. All countries must cooperate and make investments in recycling tools while educating people to change their behaviour for longer use of devices and reuse.

1.8 GLOBAL TRENDS AND POLICY IMPLICATIONS

1.8.1 Governmental and International Policies on Sustainable Tech

There is emergence of a belief that not only are the digital technologies part of the cause of climate change, but they can also offer the solution and as such the new policy path to sustainable digital innovation is evolving at cracked speed. There is enhancement in the use of digital innovation by the governments in national sustainability, improvements to laws that support green innovation, exemplary energy efficiency, and proper handling of e-waste. Such components of EU's green deal can be enumerated as the digital policy instruments for carbon neutral digital infrastructure, the circular electronics. Some of the countries that have developed national AI strategies include South Korea and Germany where sustainability is at the core have been developed. They are being stated at the global level by the United Nations World Bank and the International Telecommunication Union so that cross border innovation could be implemented, and technology could be taken to the

Global South and financed. In turn, integration of digital solutions for achievement of climate targets is explicitly acknowledged in the text of the Paris Agreement and countries are encouraged to spend money on them in the context of climatic adjustments. Still, issues of digital rights, an aspect of environmental regulation and economic development are incoherent and are in conflict with existing policies. To ensure that certain innovative digitalization can be as effective on an international level and for the poor as for developed countries, it is crucial to have international coordination as well as mandatory and identical international standards for testing and certification, and efficient mechanisms for transfer of the technologies among countries (Androniceanu, 2023; Mahardhani, 2023).

1.8.2 Corporate Social Responsibility (CSR) and Compliance

CSR that was earlier acknowledged as a marketing tool has now turned into strategic business management advantage, especially when it comes to sustainability and digitisation of business. Currently, investors, regulators, and consumers demand more significant attention to ESG issues in the company's operations than ever before. This comprises emission of carbon, supply chain sourcing, minimizing wastage on digital items, and the use of responsible AI. Some firms are very influential especially technological firms; therefore, they must adhere to most expectations concerning the use of natural resources. Many firms are preparing and releasing sustainable reports and setting specific emission reduction targets while adopting clean power sources or circular economy frameworks. Requirements for performing business across the border are constantly increasing, thus obligatory implementation of standards as ISO 14001 (environmental standard) and ISO 26000 (social responsibility). Still, the effectiveness of CSR is concomitant to the practical application of CSR. Greenwashing, which is the act of using false or exaggerated claims of company efforts in sustainable practice, is still a real threat. There should be clear and comprehensive reporting of CSR policies, engagement of the third party in the audit of CSR policies and functions of firms, and the involvement of stakeholders in determining the CSR strategies of firms to make them accountable and have meaningful impacts on sustainable digital innovation (Zhao et al., 2023).

1.8.3 Funding and Incentives for Sustainable Innovation

Financial support is one of the keys to scale up the transposition of sustainable digital innovation. Currently, there is a government-led shift in public and private financing towards green technology companies as well as renewable energy projects and climate-ready, digitally focused technologies. There are governments promoting the R&Ds in various fields, including low power computing try to provide tax credits, grants, and subsidies. Green bonds and sustainability-linked loans are examples of bonds that increase interest of investors, which seeks to finance projects with such proved benefits. At the venture capital level, there is the shift of capital to funds supporting the green economy by funding startups that are built with sustainability in mind. Development agencies are also featuring their part by providing financial supports to practices of digital sustainability in the LMICs through public-private partnerships that have the capability to deploy technical knowledge in delivering widespread improvements. Nevertheless, even today there are large gaps in funding, or in other words, angry the Internet connection is slow, the lack of adequate financing of such projects and the focus on areas where the necessary infrastructure is still being built or where the threats to the environment are greatest. Addressing such gaps will require innovative sources of financing, higher risk acceptance level of investors, and ensuring policy coherence with market incentives.

1.9 CONCLUSION

1.9.1 Recap of Emerging Trends and Technologies

Sustainability and digital innovation are two of the biggest keywords that characterize one of the biggest change processes of present days. In all fields and locations, emerging technologies are being reinvented and repositioned to address today's challenges surrounding sustainability and sustainability alone. AI is enhancing the energy technologies and making agriculture efficient and smart. IoT is giving capabilities of monitoring the environment and also the efficient use of resources in real time Edge computing concerns how processing tasks decrease the environmental load when it comes to data processing. Altogether, these technologies are not only conducive to the achievement of the sustainable development goals but are also reshaping the concept of progress for societies. But there are always two sides to the coin that cannot be denied, who lightened up the dimension of all these advances. Though digital effectively requires a significant amount of energy, leading to increased e-waste, AI is not immune to being flawed with bias, and while data supported sustainability presents a positivism, it involves a problematic aspect of privacy violation. Therefore, to apply technology and sustainability, there is the need to base it on ethical, comprehensive, and contextual approaches to cater to the long-term effects to match the short-term gains.

1.9.2 Future Directions and the Road Ahead for Sustainable Digital Innovation

As for the future of digital innovation, it must be stated that the way to follow to continue paving the road towards technical progress is and will be multilateral. Policies should be embedded within technologically progressive frameworks, society, and governments must take a central role in development of such policies, and efforts should also be coordinated internationally. In order for architecture and design to include sustainability alongside the technical and aesthetic achievements of the future, education systems must adapt to its needs – and to the need for persons who can seamlessly incorporate sustainability into their creations. This means that organisations must transition from concepts like extracting value and continued growth at the expense of stocks of resources and capital. People on their part, must play an active role in controlling their respective destinies, as well as that of their society, and be part of defining the type of digital society that is desirable and one that they wish to cultivate. ty and efficiency are now and must be more than just present in various applications of technology but also in the frameworks and the base of chips and servers, in the algorithms' architecture and the legal and regulative framework. Applying this approach contains the digitization of change for society and turns it into innovation for the improvement, sustainability, and wellbeing of all people. It is currently possible to have innovate which can be both digital and sustainable.

REFERENCES

Abad-Segura, E. et al (2020) "Sustainable management of digital transformation in higher education: Global research trends," Sustainability, 12(5), p. 2107. Available at: https://doi.org/10.3390/su12052107

Afeltra, G., Alerasoul, S.A. and Strozzi, F. (2023) "The evolution of sustainable innovation: From the past to the future," European journal of innovation management, 26(2), pp. 386–421. Available at: https://doi.org/10.1108/ejim-02-2021-0113

Androniceanu, A. (2023) "The new trends of digital transformation and artificial intelligence in public administration," ADMINISTRATIE SI MANAGEMENT PUBLIC, (40), pp. 147–155. Available at: https://www.ceeol.com/search/article-detail?id=1231420

Geng, Y., Zhang, N. and Zhu, R. (2023) "Research progress analysis of sustainable smart grid based on CiteSpace," Energy strategy reviews, 48(101111), p. 101111. Available at: https://doi.org/10.1016/j.esr.2023.101111

Ghobakhloo, M. et al. (2021) "Industry 4.0, innovation, and sustainable development: A systematic review and a roadmap to sustainable innovation," Business strategy and the environment, 30(8), pp. 4237–4257. Available at: https://doi.org/10.1002/bse.2867

Hassan, Q. et al. (2024) "Enhancing smart grid integrated renewable distributed generation capacities: Implications for sustainable energy transformation," Sustainable energy technologies and assessments, 66(103793), p. 103793. Available at: https://doi.org/10.1016/j.seta.2024.103793

Hrustek, L. (2020) "Sustainability driven by agriculture through digital transformation," Sustainability, 12(20), p. 8596. Available at: https://doi.org/10.3390/su12208596

Hu, H., Qi, S. and Chen, Y. (2023) "Using green technology for a better tomorrow: How enterprises and government utilize the carbon trading system and incentive policies," China economic review, (101933), p. 101933. Available at: https://doi.org/10.1016/j.chieco.2023.101933

Kabeyi, M.J.B. and Olanrewaju, O.A. (2023) "Smart grid technologies and application in the sustainable energy transition: A review," International journal of sustainable energy, 42(1), pp. 685–758. Available at: https://doi.org/10.1080/14786451.2023.2222298

Lastman (2025b) Kaggle.com. Available at: https://www.kaggle.com/datasets/lastman0800/future-trends-in-digital-innovation

Liu, X., Liu, F. and Ren, X. (2023) "Firms' digitalization in manufacturing and the structure and direction of green innovation," Journal of environmental management, 335, p. 117525. Available at: https://doi.org/10.1016/j.jenvman.2023.117525

Mahardhani, A.J. (2023) "The role of public policy in fostering technological innovation and sustainability," Journal of contemporary administration and management (ADMAN), 1(2), pp. 47–53. Available at: https://doi.org/10.61100/adman.v1i2.22

Omol, E.J. (2024) "Organizational digital transformation: From evolution to future trends," Digital transformation and society, 3(3), pp. 240–256. Available at: https://doi.org/10.1108/dts-08-2023-0061

Salvi, M. et al (2022) Towards a green & digital future: Key requirements for successful twin transitions in the European Union. Publications Office of the European Union.

Satalkina, L. and Steiner, G. (2020) "Digital entrepreneurship and its role in innovation systems: A systematic literature review as a basis for future research avenues for sustainable transitions," Sustainability, 12(7), p. 2764. Available at: https://doi.org/10.3390/su12072764

Trevisan, L.V. et al. (2023) "Digital transformation towards sustainability in higher education: State-of-the-art and future research insights," Environment development and sustainability, 26(2), pp. 1–22. Available at: https://doi.org/10.1007/s10668-022-02874-7

Yin, S., Liu, L. and Mahmood, T. (2023) "New trends in sustainable development for industry 5.0: Digital green innovation economy," Green and low-carbon economy, 2(4), pp. 269–276. Available at: https://doi.org/10.47852/bonviewglce32021584

Zhao, Q., Li, X. and Li, S. (2023) "Analyzing the relationship between digital transformation strategy and ESG performance in large manufacturing enterprises: The mediating role of green innovation," Sustainability, 15(13), p. 9998. Available at: https://doi.org/10.3390/su15139998

2 The Impact of ESG Reporting on Corporate Strategy and Consumer Trust

A Systematic Literature Review

Shad Ahmad Khan and Yogesh Sinha

2.1 INTRODUCTION

Environmental, Social, and Governance (ESG) models have emerged as primary tools for assessing corporate responsibility and long-term value creation, notwithstanding there is little known about the direct influence of ESG reporting on corporate strategy (CS) and consumer trust (CT). As much as many organizations are adopting ESG practices, reporting standards variations, stakeholder requirements variations, and the empirical reality of unclear tangible benefits resulting from ESG openness pose enormous challenges. Further, with younger consumers and generations showing a preference for sustainable and ethical consumption practices, firms are under mounting pressure not only to adopt ESG initiatives but also to credibly report them. While ESG is becoming increasingly important, there are disconnected scholarly papers with very few all-encompassing reviews connecting the disclosure of ESG with strategic fit and consumer behavior. This research fills this important knowledge gap through critical analysis of current research to guide how ESG disclosure affects CS and CT. The aim of the study is to elucidate the way in which ESG transparency can be used to build stakeholder trust, brand reputation, and help firms balance the delicate equilibrium between profitability and sustainability.

ESG frameworks are structured approaches by which organizations monitor and report their performance beyond the conventional financial measures. Principles of ESG frameworks have become extremely significant in shaping corporate behavior, stakeholder expectations, and regulation. The environmental aspect evaluates a firm's impact on natural environments; the social one pertains to engagement with employees, customers, and society; and governance pertains to management, ethics, and transparency in corporate behavior. Various global frameworks—including the Global Reporting Initiative (GRI), Sustainability Accounting Standards Board (SASB), and Task Force on Climate-related Financial Disclosures (TCFD) have

DOI: 10.1201/9781003611905-3

seemed to regulate ESG disclosure (Chua et al., 2022). With more responsibility and transparency needed from consumers, investors, and regulators, today ESG reporting is no longer an extras nice-to-have task but a strategic must-have for businesses these days (Kandpal et al., 2024).

Incorporating ESG reporting into CS is what certifies the shift from shareholder capitalism to stakeholder capitalism. Firms must create long-term value for investors but also for society as a whole. ESG reporting is at the center of linking CS with sustainability and ethics. At the same time, consumers are becoming increasingly aware of corporate social responsibility, seeking greater transparency and honesty from brands they consume. Trust, which was once determined largely by the product quality or value, today increasingly is being influenced by how companies go about their social and environmental responsibilities. Learning about ESG disclosure from this dual perspective—that is, strategic corporate alignment and customer trust—teaches us an important lesson on how the firms navigate the tightrope between profitability and sustainability amidst an increasingly complex global setting.

This chapter addresses a timely issue. The unprecedented acceleration of global warming, escalating social inequalities, and recent company scandals have accelerated the need for trustworthy, open, and relevant ESG practice. Furthermore, regulatory innovation such as the European Union's (EU) Corporate Sustainability Reporting Directive (CSRD) and increasing integration of ESG factors in investment decisions have increased the call for corporations to adopt ESG principles in their strategies. At the same time, the emergence of ethically aware consumers, particularly in cohorts like Gen Z, has resulted in CT becoming a central success driver for brands. Systematic review of current know-how is therefore important at this point to integrate knowledge, plot holes, and inform future practice and research. The primary objective of this chapter is to critically review the existing scholarly literature which responds to the impact of the ESG report on CS and CT. The study aims to assess the relationship between the transparency of the ESG and CT, and outlines an international picture of the ESG, CS, and CT.

2.2 METHODOLOGY

This chapter uses the systematic literature review (SLR) approach to investigate available scholarly literature at the intersection of ESG reporting, CS, and CT. The Scopus database was utilized as the primary source of literature identification because it provides wide coverage of peer-reviewed journals in business, management, sustainability, and social sciences disciplines. Scopus has been famously known for its strict indexing standards, range of global journals, and nuanced search functions, such as the option for exact filtering by keywords, document types, time frames, and subject areas (Oliveira et al., 2018). The research focused on the 2020–2025-time frame to determine the evolution of ESG practices during a significant era marked by increasing regulatory attention and stakeholder activism. Table 2.1 summarizes the inclusion and exclusion criteria used in this study.

TABLE 2.1
Literature Search Criteria

Search Theme	Inclusion and Exclusion Criteria	No. of Document Extracted	Documents Found as per the Theme of the Chapter
Corporate Strategy	"corporate strategy" OR "business strategy" Query Run: (TITLE-ABS-KEY("corporate strategy" OR "business strategy") AND PUBYEAR > 2019 AND PUBYEAR < 2026 AND (LIMIT-TO (EXACTKEYWORD, "Corporate Strategy") OR LIMIT-TO (EXACTKEYWORD, "Business Strategy")) AND (LIMIT-TO (LANGUAGE, "English"))	3,538	Bibliometric analysis was performed due to large document size
ESG Reporting	("ESG reporting" OR "environmental social governance" OR "sustainability reporting") Query Run: TITLE-ABS-KEY(("ESG reporting" OR "environmental social governance" OR "sustainability reporting")) AND PUBYEAR > 2019 AND PUBYEAR < 2026 AND (LIMIT-TO (EXACTKEYWORD, "Corporate Strategy") OR LIMIT-TO (EXACTKEYWORD, "Business Strategy")) AND (LIMIT-TO (LANGUAGE, "English")))	50	14 (Al Astal et al., 2025; Bhat et al., 2024; Blagov & Petrova-Savchenko, 2021; Chen et al., 2023; Diwan & Amarayil Sreeraman, 2024; Jeong et al., 2023; Jiménez-Montañes & de Gracia, 2023; Lascols, 2021; Sarkar et al., 2023; Schneider et al., 2025; Singhania & Saini, 2023; Widyawati, 2020; Zaman et al., 2021; Zampone et al., 2023)
Consumer Trust	Query Run: (TITLE-ABS-KEY(("consumer trust" OR "stakeholder trust")) AND PUBYEAR > 2019 AND PUBYEAR < 2026 AND (LIMIT-TO (EXACTKEYWORD, "Corporate Strategy") OR LIMIT-TO (EXACTKEYWORD, "Business Strategy")) AND (LIMIT-TO (LANGUAGE, "English")))	7	7 (Ammer et al., 2020; Huang et al., 2025; Konuk, 2018; Krishna et al., 2024; Tong et al., 2022; Wang et al., 2021; Yu et al., 2021)

(Continued)

TABLE 2.1 ***(Continued)***
Literature Search Criteria

Search Theme	Inclusion and Exclusion Criteria	No. of Document Extracted	Documents Found as per the Theme of the Chapter
Impact of ESG Reporting on Corporate Strategy and Consumer Trust	Query Run: (“ESG reporting” OR “environmental social governance” OR “sustainability reporting” OR “non-financial reporting”) AND (“corporate strategy” OR “strategic management” OR “business strategy” OR “corporate planning” OR “strategic alignment”) AND (“consumer trust” OR “customer trust” OR “stakeholder trust” OR “consumer perception” OR “brand trust” OR “consumer confidence”)	2	2 (Antonia Garcia-Benau et al., 2013; Brogi & Lagasio, 2025)

2.3 FINDINGS AND DISCUSSION

2.3.1 ESG and ESG Reporting

ESG considerations have been the leading aspect of corporate sustainability, and they have guided investment, stakeholders' confidence, and business sustainability in the long run. ESG reporting, a measure of the performance of a firm in ESG issues, has been a valuable means of transparency and accountability. It is confirmed by evidence that successful ESG practices can increase financial performance, minimize risk, and align business strategies with global sustainability aspirations such as the UN Sustainable Development Goals (SDGs; Chen et al., 2023; Sarkar et al., 2023). For instance, firms embedding ESG within mainstream strategy will tend to enhance brand reputation and operational efficiency, particularly in pollution-intensive industries (Chen et al., 2023).

But ESG reporting remains beset by challenges from inconsistency in measurement, standardization, and by divergent regulatory demands. Widyawati (2020) elucidates that whereas ESG metrics are mandatory in socially responsible investing (SRI), they are undercut on reliability by low transparency and divergent frameworks. Similarly, Zaman et al. (2021) emphasize the role of audit committees in improving the quality of sustainability assurance where independent monitoring and industry knowledge are recognized as critical to issuing credible reports. The application of international standards like the GRI and the SASB has emerged as the turning point in ensuring harmonized disclosures (Diwan & Amarayil Sreeraman, 2024; Singhania & Saini, 2023).

The growing emphasis on ESG is a shift from traditional financial reporting to more expansive stakeholder-based thinking. Companies are now required to demonstrate their role in environmental protection, social justice, and ethical leadership as opposed to profitability (Bhat et al., 2024; Schneider et al., 2025). With regulatory forces mounting—caused by the EU's CSRD—companies must combine ESG disclosures with materiality considerations to avoid greenwashing and build long-term credibility (Jiménez-Montañes & de Gracia, 2023).

Schneider et al. (2025) found that companies with strongly embedded sustainability strategies, that is, those most closely linked to core business models and overseen by senior management, that is, report superior bottom-line financial performance, in the form of return on assets. Blagov and Petrova-Savchenko (2021) also argue the need to shift from traditional corporate social responsibility to a "shared value" approach integrating ESG into a company's competitive strategy.

On the other hand, there are firms that may employ ESG reporting for symbolic rather than substantive purposes. Zampone et al. (2023) caution that peer rivalry and mimicry tend to result in superficial SDG reporting, thereby undermining its credibility. Lascols (2021) proposes a novel framework for accounting externalities with a view to strengthening ESG's accountability through the introduction of new financial reports conveying social and environmental impacts.

Jeong et al. (2023) highlight regional differences in the ESG performance of companies worldwide. Their cross-national comparison of Korean biopharmaceutical companies shows decoupling between financial performance and ESG scores, suggesting sectoral impediments to the adoption of ESG. Al Astal et al. (2025) follow up by advocating for the use of specially tailored management control systems (MCS) to execute ESG strategies and map internal governance to sustainability goals.

2.3.2 Corporate Strategy

Due to the large document size extracted (3,538), the need was felt to retain the documents as the main objective of the extraction was to understand the concept the developments in recent years in the context of CS. CS has been the underlying factor from the corporate success (Khan & Ansari, 2025). CS is the total, long-range plan that expresses a company's vision, direction, and scope to achieve great goals and create value. CS assists the top management to make decisions about which markets to enter, how to allocate resources among business units, and how to sustain competitive advantage in the industry. CS is effectively a roadmap or game plan of the company's future, orchestrating all the organizational efforts to growth, profitability, and ultimate success (Bubenik et al., 2022; Paley, 2021).

In order to understand the recent development in the field of CS, bibliometric analysis was performed utilizing vos-viewer (Magd et al., 2023; Saxena et al., 2024; Sharma et al., 2024). In order to understand this co-occurrence analysis was performed where 30 keywords were extracted by keeping the minimum occurrence threshold as 108. Table 2.2 presents the same.

The researchers further applied the human intelligence to draw meaningful conclusions from the above extracted keywords by removing the keywords that were not aligned with the objectives of this study, a total of 10 keywords were extracted by this approach. The refined Table 2.3 is presented with the interpretation column.

There were few emerging themes that were aligned with the theme of this chapter as presented in Table 2.4.

The bibliometric co-occurrence of CS research reveals a dynamic and multidimensional discipline in which the main topic of CS is inseparably intertwined with such topics as business strategy, governance, innovation, and sustainability. The frequency and strength of association of such terms as business strategy, corporate governance, sustainable development, and innovation indicate an acute scholarly emphasis on aligning strategic objectives with long-term environmental and social agendas. The prominence of China as a contextual keyword suggests a regional emphasis of research, possibly driven by its rapid economic transformation and strategic policy development. In addition, the inclusion of terms like digital transformation, climate change, and stakeholders suggests that CS is increasingly driven by technological advancement, environmental concerns, and expanded stakeholder expectations.

The frequent and deep interlinking of terms such as sustainability, corporate governance, stakeholder, climate change, and corporate social responsibility shows the growing importance of ESG issues in strategic business decision-making. The infusion of environmental economics and governance philosophy

TABLE 2.2
Co-occurrence Keywords Extracted from Vos-Viewer

	Keyword	Occurrences	Total Link Strength
1.	Article	128	599
2.	Business	114	343
3.	Business development	207	670
4.	Business strategy	1,322	1,322
5.	China	568	1,877
6.	Climate change	110	373
7.	Commerce	143	414
8.	Competition	109	285
9.	Corporate governance	263	822
10.	Corporate social responsibility	126	396
11.	Corporate strategy	2,266	4,719
12.	Corporates	125	496
13.	Decision-making	205	577
14.	Digital transformation	137	350
15.	Empirical analysis	116	360
16.	Environmental economics	276	1,102
17.	Finance	117	398
18.	Governance approach	309	1,041
19.	Human	111	496
20.	Industrial performance	231	813
21.	Innovation	440	1,387
22.	Investment	212	669
23.	Management practice	108	338
24.	Manufacturing	138	468
25.	Performance assessment	133	503
26.	Stakeholder	181	623
27.	Strategic approach	134	507
28.	Strategic planning	224	475
29.	Sustainability	480	1,584
30.	Sustainable development	371	1,327

in the strategic mindset underscores how ESG reporting is shaping open, transparent, and forward-looking strategies in harmony with shifting stakeholder expectations. Furthermore, the increasing dominance of CT is also suggested by performance measurement, innovation, and CSR themes, which indicate that firms using ESG reporting not only align their strategies with ethical and sustainable values but also establish credibility and trust with more conscious consumers. This intersection of ESG practices and business strategy is a paradigm shift from profit models to value-based strategies creating long-term brand equity and stakeholder affinity.

TABLE 2.3
Refined Keywords Extracted for CS

Keyword	Occurrences	Total Link Strength	Interpretation
Corporate Strategy	2,266	4,719	Central theme of the study—most connected and frequently occurring term.
Business Strategy	1,322	1,322	Closely tied with corporate strategy; high relevance.
China	568	1,877	Suggests geographic focus—numerous studies on corporate strategy involve Chinese firms or contexts.
Innovation	440	1,387	A key factor in corporate strategy, reflecting themes of competitiveness and adaptability.
Sustainability	480	1,584	It indicates a strong connection between sustainability and corporate strategy.
Sustainable Development	371	1,327	Closely related to above, showing strategic focus on long-term goals.
Corporate Governance	263	822	Strong presence: governance is a major pillar in strategy discussions.
Governance Approach	309	1,041	Reinforces above, likely linked with ethical, responsible decision-making.
Environmental Economics	276	1,102	Reflects integration of ecological perspectives into strategic thinking.
Industrial Performance	231	813	Suggests measurement and evaluation of strategy outcomes.

TABLE 2.4
Emerging Themes Identified for CS

Keyword	Occurrences	Total Link Strength	Interpretation
Digital Transformation	137	350	Modern strategic shifts in tech-oriented firms.
Climate Change	110	373	Environmental challenges influencing corporate decisions.
Corporate Social Responsibility (CSR)	126	396	A vital element of strategy in sustainability discourse.
Stakeholder	181	623	Reinforces stakeholder theory in strategy.
Strategic Planning/ Strategic Approach	224/134	475/507	Tools and techniques in strategy execution.

2.3.3 Understanding Consumer Trust and Its Contextual Relation to ESG

CT plays a dominant role in purchasing behavior, loyalty, and longer-term stakeholder relationships. Trust tends to be fostered by transparency, good ethics, and reliable honoring of commitments, values closely integrated with ESG practice. Fresh research provides insightful information regarding the dynamic relationship between CT and ESG considerations.

Yu et al. (2021) reported that CSR image and corporate capability enhance CT in organic food firms. CSR image, interestingly, played a more significant role than corporate capability image, which implies that ethical and sustainable conduct is of greater significance in generating trust compared to functional ability per se. This result fits within the "Social" and "Governance" pillars of ESG that show how ethical conduct generates CT.

Tong et al. (2022) further examined the influence of negative publicity on trust and found that negative publicity destroys brand trust via diminished image congruity, that is, functional as well as self-congruity. They found that efforts to restore include compensation and public apology, though whose efficacy resided in the content of the adverse publicity, since being performance versus value based. This underscores the importance of advance ESG disclosure and transparency to counter trust erosion during crises through early response to stakeholder concerns.

Technological advancements, particularly blockchain, have also been observed to be promising toward the establishment of CT. Krishna et al. (2024) demonstrated that blockchain technology improves supply chain transparency, therefore reducing fraud and information asymmetry. This, consequently, maximizes stakeholder trust by maximizing data security and traceability. Blockchain facilitates ESG aspirations through enabling ethical sourcing (Environmental) and fair labor practices (Social), while its decentralized nature enhances Governance ideals.

In contrast, Huang et al. (2025) showed that greenwashing perceptions, wherein companies exaggerate or make false claims regarding environmental involvement, disintegrate customer trust, and impair perceived environmental performance. However, firms with positive environmental policies and high embeddedness in their networks were able to reduce perceptions of greenwashing. This is proof of the relevance of authentic ESG activities, such as accredited sustainability reports, in hindering greenwashing and maintaining customer trust.

Finally, Wang et al. (2021) noted the role of social media in transferring trust, that CT in social media brand communities can be transferred to the brand itself, ultimately influencing purchase intentions and word-of-mouth. Such brands as can leverage the influence of ESG stories on social media, say, by emphasizing sustainability stories that can enhance trust and engagement with the audience.

In general, the literature demonstrates that ESG-aligned approaches not only satisfy regulatory and ethical demands but also serve as a platform from which to establish and sustain consumer confidence. However, the literature provides few more details when it is checked from the lens of ESG and ESG reporting.

CT and ESG share an inextricably contextual and mutually sustaining relationship. Transparency serves as a primary bridge between the two. ESG disclosures such as sustainability reports are critical in the development of trust because they present firm commitments that are verifiable (Ammer et al., 2020). The application of blockchain technology in ESG reporting reinforces this trust even more with unalterable and tamper-proof records, thereby dispelling consumer skepticism regarding the authenticity of sustainability claims (Krishna et al., 2024).

Ethical consumerism has also been a major driver of trust, as consumers increasingly make purchasing decisions based on corporations' ethical conduct. Research in the organic food sector points to the way CT is directly connected to ESG-focused traits such as fair trade, green manufacturing, and social responsibility (Konuk, 2018; Yu et al., 2021). This identifies heightened consumer demand for brands aligned with their values.

In times of crisis, firms that have strong ESG frameworks prove to be more resilient and gain back confidence quickly. Tong et al. (2022) determined that companies with strong ESG frameworks can better manage reputation risks, in other words, ESG interventions not only work as shields but also as handy recovery mechanisms when there is social unrest. Apart from that, stakeholder engagement is another support pillar of ESG-oriented companies. Particularly in industries like organic food, organizations that interact with stakeholders collaboratively and emphasize inclusiveness and shared values are more likely to build long-term trust (Yu et al., 2021).

The policy implications in real-world terms are considerable. For business, integration of ESG values into core activities is critical to winning the hearts of true CT, and blockchain represents a useful technology to enhance ESG transparency. Policymakers have a key contribution to make through imposing uniformed ESG disclosures, which can help curtail greenwashing and engender trust within whole industries. For researchers, there is fertile ground to cultivate the way that regional and cultural considerations shape the trust—ESG relationship, particularly the distinction between emerging and developed markets.

Hence, CT is both an outcome of and cause of ESG excellence. Companies that align ESG activity with consumer expectations not only reduce reputational and operational risks but also gain a competitive edge with the building of loyal, activated stakeholders.

2.3.4 Impact of ESG Reporting on Corporate Strategy and Consumer Trust

There are only two papers identified on the Scopus database that have dealt with the intended keywords, that is, Antonia Garcia-Benau et al. (2013) and Brogi and Lagasio (2025).

The meta-analysis by Antonia Garcia-Benau et al. (2013) demonstrates that certain corporate governance features—board size and female representation—positively influence ESG disclosure and profitability, however, independent directors do not have any significant influence; further, superior quality ESG disclosure is strongly correlated with excellent financial performance, emphasizing the key role played by good governance structures in driving integrated sustainability reporting and long-term business success. This emphasizes the necessity to integrate governance and sustainability approaches to promote transparency, accountability, and stakeholder confidence, an observation also echoed by related studies citing the need for compulsory governance disclosures along with environmental and social data to support effective ESG risk management and shareholder accountability.

The study by Brogi and Lagasio (2025) primarily deals in the arena of financial crisis where the study finds that, even in the time of financial crisis, Spanish companies strengthened CSR reporting and maintained stable assurance practices, which suggests that CSR disclosure and assurance are strategic investments that enhance corporate performance and stakeholder confidence. Although Brogi and Lagasio (2025) discovered that there were no actual changes in CSR assurance practices during crises, cross-country research upholds that third-party assurance enhances ESG credibility. CSR-assured firms, for example, enjoy higher market valuation, particularly in countries with low financial reporting standards. This can be interpreted as assurance coming across as an investment in transparency in the long term rather than a cost-saving goal, facilitating strategic differentiation. Further, the study suggested that ESG disclosure becomes more embedded in governance frameworks to address investors' needs.

Companies applying ESG transparency attract sustainable capital and reduce capital costs. For example, businesses with high consumer proximity make ESG disclosures their top priority to align the values of customers for competitive advantage (Brogi & Lagasio, 2025). The study added that transparent reporting of ESG improves CT and loyalty by being stakeholder value-aligned and greenwashing risk-minimizing, especially in crisis moments, as those firms that transparently report and have attested to their sustainability initiatives are viewed as more credible, responsible, and dedicated, building brands and long-term customer relationships.

Following Antonia Garcia-Benau et al. (2013) and Brogi and Lagasio (2025), ESG reporting plays a strategic and trust-enhancing role in corporate governance and market positioning. Garcia-Benau et al. highlight that good governance structures, that is, appropriate board composition and high-quality ESG disclosures are key to achieving financial outperformance and enabling transparency and accountability. This is corroborated by the work of Brogi and Lagasio, wherein it is evident that even during periods of financial crises, consistent and credible ESG reporting, especially if assured by third parties, deepens CT and enhances stakeholder confidence. Taking together, these writings underscore that ESG reporting is not an exercise in compliance but a strategic imperative that involves governance, sustainability, and stakeholder engagement to build resilient, trusted, and competitively differentiated enterprises.

2.4 CONCLUSION

This systematic review tried to synthesize evidence on the impact of ESG reporting on CS and CT. The review supported the fact that ESG guidelines have been at the heart of tracking and reporting on corporate performance, alongside financial considerations. ESG reporting is increasingly being viewed as a strategic priority, essential in the alignment of business strategy and sustainability and ethics in an environment requiring more corporate responsibility. The study highlights that effective ESG practices can enhance financial performance, risk reduction, and global sustainability goals. However, the landscape of ESG reporting is plagued by inconsistency in metrics and competing regulatory expectations, highlighting the utmost need for harmonized disclosure through standards like GRI and SASB.

The review determines that CS, as the roadmap for a company's future, is evolving. Current research indicates that CS is increasingly integrated with sustainability, governance, innovation, and stakeholder concerns. This suggests a paradigm shift away from purely profit-maximizing models toward value-based strategies with the goal of creating long-term brand equity and stakeholder attachment.

Moreover, the research depicts the pivotal role of CT, which is highly correlated with perceived corporate ethical conduct and transparency. Evidence shows that sustainable and ethical conduct, in line with the Social and Governance pillars of ESG, have significant impact in building CT, sometimes more than mere functional competence. Technologies like blockchain have been considered to be promising tools to enhance transparency and verify sustainability statements, hence boosting trust and avoiding greenwashing. Companies with robust ESG frameworks appear to be in a better position to manage reputation risk and regain confidence during crises.

While there is extensive literature on ESG, CS, and CT in general, only two studies addressing the impact of ESG reporting on CS and CT were found by the systematic search. The limited number of studies suggests that good-governance arrangements and guaranteed high-quality ESG disclosures have a positive impact on financial performance, enhance transparency, and build stakeholder confidence, including CT, especially in periods of adversity. This underscores that good ESG reporting is not merely a compliance issue but a strategic investment in transparency, credibility, and long-term resilience. In summary, the literature surveyed identifies the dynamic development of CS toward the integration of ESG considerations and the key role of transparent ESG reporting in establishing and maintaining consumer and stakeholder trust.

2.5 IMPLICATIONS OF THE STUDY

The findings of this systematic review have far-reaching implications for stakeholders in general. On the Managerial front the companies need to prioritize transparency and accuracy in their ESG reporting. The new tools like sustainability reports and blockchain technology provide verifiable data and eliminate greenwashing doubts. Further, investment in good governance and third-party assurance of ESG reporting can enhance credibility and resilience. Having ESG as a focus can significantly influence CT. Further, this study is an implication for the policymakers and the regulatory organizations like the EU's CSRD that are compelling the companies to combine ESG disclosures with materiality considerations.

The relative lack of studies on how ESG reporting impacts both CT and CS shows significant literature deficit. There is fertile ground for potential future research on exploring this very intersection further. Further, the need to consider how cultural and regional issues bear on the relationship between trust and ESG, particularly comparing developed and emerging markets, provides an important line for future research.

Overall, implications indicate a shift toward a new paradigm where the pressure is forcing companies to marry profitability with sustainability and ethics through mounting regulatory enforcement and demands for ethically aware consumers. Adoption of ESG thinking strategically along with open and verifiable disclosure reporting are the ways to come out successful within such an ecosystem.

2.6 DIRECTION FOR FUTURE RESEARCH

Regarding the gaps and themes of the literature identified, several future research directions are proposed in order to better explain the complex interplay among ESG reporting, CS, and CT. Future research needs to explicitly examine the cumulative impact of ESG reporting on CS and CT with both quantitative and qualitative research approaches. It is also crucial to explore regional and cultural nuances by comparing emerging and mature markets to understand stakeholder expectations and the success of ESG communication. A further exploration of mechanisms of corporate governance and third-party assurance involvement can offer insight into how these drivers shape ESG credibility and impact. The use of technologies like blockchain needs to be explored to assess whether they have the potential to enhance ESG transparency and trust. Besides, scientists need to conduct component-specific analysis to determine the independent effects of the ESG pillars. An essential field of concern should be distinguishing real ESG practices and greenwashing, and their respective perception and reaction by customers. Finally, longitudinal studies are required to track THE long-term course of corporate responses to tackling ESG integration and measure the long-term impact of transparent ESG reporting on consumer confidence. Together, these research directions underscore the need for more stringent, precise, and more context-specific inquiry.

REFERENCES

Al Astal, A. Y. M., Alzoraiki, M., Ateeq, A., Milhem, M., Ateeq, R. A., & Santhanamery, T. (2025). Enhancing ESG implementation through effective management control systems. In J. Kacprzyk (Ed.), *Business Sustainability with Artificial Intelligence (AI): Challenges and Opportunities* (pp. 647–656). Springer, Cham.

Ammer, M. A., Aliedan, M. M., & Alyahya, M. A. (2020). Do corporate environmental sustainability practices influence firm value? The role of independent directors: Evidence from Saudi Arabia. *Sustainability*, *12*(22), 9768.

Antonia Garcia-Benau, M., Sierra-Garcia, L., & Zorio, A. (2013). Financial crisis impact on sustainability reporting. *Management Decision*, *51*(7), 1528–1542.

Bhat, B. A., Makkar, M. K., & Gupta, N. (2024). Corporate governance for sustainability: Connecting environmental, social and economic goals. *Sustainable Development*, *32*(3), 2754–2765.

Blagov, Y. E., & Petrova-Savchenko, A. A. (2021). The transformation of corporate sustainability model in the context of achieving the UN SDGs: Evidence from the leading Russian companies. *Corporate Governance: The International Journal of Business in Society*, *21*(2), 307–321.

Brogi, M., & Lagasio, V. (2025). Do corporate governance frameworks affect sustainability reporting? *Corporate Social Responsibility and Environmental Management*, *32*(3), 3928–3943.

Bubenik, P., Capek, J., Rakyta, M., Binasova, V., & Staffenova, K. (2022). Impact of strategy change on business process management. *Sustainability, 14*(17), 11112.

Chen, S., Song, Y., & Gao, P. (2023). Environmental, social, and governance (ESG) performance and financial outcomes: Analyzing the impact of ESG on financial performance. *Journal of Environmental Management, 345*, 118829.

Chua, W. F., James, R., King, A., Lee, E., & Soderstrom, N. (2022). Task Force on Climate-related Financial Disclosures (TCFD) implementation: An overview and insights from the Australian accounting standards board dialogue series. *Australian Accounting Review, 32*(3), 396–405.

Diwan, H., & Amarayil Sreeraman, B. (2024). From financial reporting to ESG reporting: A bibliometric analysis of the evolution in corporate sustainability disclosures. *Environment, Development and Sustainability, 26*(6), 13769–13805.

Huang, F., Wu, J., Wu, Z., Fu, W., Guo, P., Zhang, Z., & Khan, F. (2025). Unpacking greenwashing: The impact of environmental attitude, proactive strategies, and network embeddedness on corporate environmental performance. *Journal of Environmental Management, 373*, 123625.

Jeong, H., Song, C., Kim, S., & Shin, K. (2023). Quantum jump in biopharmaceutical industry: A case of Korea's catching up with Europe and US. *European Planning Studies, 31*(5), 885–904.

Jiménez-Montañes, M. A., & de Gracia, S. V. (2023, July). Sustainability as an Informative Business Strategy: New EU Directive and the Principle of Double Materiality. In J. Kacprzyk (Ed.), *International Symposium on Distributed Computing and Artificial Intelligence* (pp. 88–98). Springer Nature Switzerland, Cham.

Kandpal, V., Jaswal, A., Santibanez Gonzalez, E. D., & Agarwal, N. (2024). Corporate social responsibility (CSR) and ESG reporting: Redefining business in the twenty-first century. In A. Stefanakis & I. Nikolaou (Eds.), *Energy Transition: Circular Economy and Sustainable Financing for Environmental, Social and Governance (ESG) Practices* (pp. 239–272). Springer Nature Switzerland, Cham.

Khan, S. A., & Ansari, M. S. A. (2025). Critical factors for relationship between strategic orientation, corporate success, and innovation. In *Human Perspectives of Industry 4.0 Organizations* (pp. 89–110). CRC Press. Boca Raton, FL.

Konuk, F. A. (2018). The role of store image, perceived quality, trust and perceived value in predicting consumers' purchase intentions towards organic private label food. *Journal of retailing and consumer services, 43*, 304–310.

Krishna, I. M., Prabha, R., Shravani, M., Sri, A. S., Tejaswini, V., & Vyshnavi, T. S. (2024, March). A business strategy for ensuring trust and transparency in supply chain management using blockchain based data security. In *2024 2nd International Conference on Disruptive Technologies (ICDT)* (pp. 121–127). IEEE.

Lascols, B. (2021). Accounting and reporting for externalities: Balance of externalities. *Sustainability and Climate Change, 14*(2), 115–121.

Magd, H., Khan, S. A., Khatri, B., Sharma, N., & Arora, S. (2023). Understanding the relationship between IoT and digital marketing: A bibliometric analysis. In A. Naim & V. A. Dev (Eds.), *Global Applications of the Internet of Things in Digital Marketing* (pp. 123–140). IGI Global, Hershey, PA.

Oliveira, A. S., de Barros, M. D., de Carvalho Pereira, F., Gomes, C. F. S., & Da Costa, H. G. (2018). Prospective scenarios: A literature review on the Scopus database. *Futures, 100*, 20–33.

Paley, N. (2021). *The manager's guide to competitive marketing strategies.* Routledge, London.

Sarkar, S., Moolearambil Sukumaran Nair, M., & Datta, A. (2023). Role of environmental, social, and governance in achieving the UN sustainable development goals: A special focus on India. *Environmental Progress & Sustainable Energy, 42*(6), e14204.

Saxena, C., Khatri, B., & Khan, S. A. (2024). Mapping the landscape of social sustainability research: A bibliometric analysis and sentiment analysis. In R. A. Castanho (Ed.), *ESG and Ecosystem Services for Sustainability* (pp. 309–336). IGI Global Scientific Publishing, Hershey, PA.

Schneider, A., Kayser, C., Retsch, B. T., Thun, T. W., & Zülch, H. (2025). Sustainability integration matters! A German perspective on financial top and bottom line performance. *Journal of cleaner production*, *501*, 145309.

Sharma, N., Khatri, B., & Khan, S. A. (2024). Exploring the past and identifying the future research agenda for social media and tourist behavior: A bibliometric analysis. In C. M. Q. Ramos, T. Costa, F. S. Severino, and M. de Lurdes Calisto (Eds.), *Social Media Strategies for Tourism Interactivity* (pp. 261–284).

Singhania, M., & Saini, N. (2023). Institutional framework of ESG disclosures: Comparative analysis of developed and developing countries. *Journal of Sustainable Finance & Investment*, *13*(1), 516–559.

Tong, Z., Feng, J., & Liu, F. (2022). Understanding damage to and reparation of brand trust: A closer look at image congruity in the context of negative publicity. *Journal of Product & Brand Management*, *32*(1), 157–170.

Wang, X., Wang, Y., Lin, X., & Abdullat, A. (2021). The dual concept of consumer value in social media brand community: A trust transfer perspective. *International Journal of Information Management*, *59*, 102319.

Widyawati, L. (2020). A systematic literature review of socially responsible investment and environmental social governance metrics. *Business Strategy and the Environment*, *29*(2), 619–637.

Yu, W., Han, X., Ding, L., & He, M. (2021). Organic food corporate image and customer co-developing behavior: The mediating role of consumer trust and purchase intention. *Journal of Retailing and Consumer Services*, *59*, 102377.

Zaman, R., Farooq, M. B., Khalid, F., & Mahmood, Z. (2021). Examining the extent of and determinants for sustainability assurance quality: The role of audit committees. *Business Strategy and the Environment*, *30*(7), 2887–2906.

Zampone, G., García-Sánchez, I. M., & Sannino, G. (2023). Imitation is the sincerest form of institutionalization: Understanding the effects of imitation and competitive pressures on the reporting of sustainable development goals in an international context. *Business Strategy and the Environment*, *32*(7), 4119–4142.

Part II

Digital Transformation and Industry 4.0

3 Challenges and Opportunities of IR 4.0 for Financial Institutions

Shoaib Irshad and Muzammil Khurshid

3.1 INTRODUCTION

The unification of Industry 4.0 (IR 4.0) in the field of finance is making things more secure, productive, personalized and useful, especially in delivering financial services. The evolution of IR 4.0 has brought several revolutionary technologies into financial services, like artificial intelligence (AI), which facilitates risk management and investment strategies, blockchain for transparency, Internet of Things (IoT) for asset monitoring and quantum computing for data. These technologies fulfill the needs and expectations of the customers, as well as regulatory problems and technical requirements (Hasan et al., 2024). IR 4.0 has multi-dimensional effects in the field of financial services. It is a tool to create a new range of financial products and services, such as digital banking and AI-generated investment channels, which are examples. On the other hand, it points out the important questions regarding the future, such as the wastage of skilled and competent human capital. There is a need to counter its implications for the financial sector, including its impact on financial stability, regulatory frameworks, and societal well-being.

The emergence of IR 4.0 in the financial sector has opened the new heavens to discover new platforms through which financial institutions can grow rapidly. Such technologies have improved the performance of financial institutions by ensuring secure risk management supersystems, enhancing the efficiency in operational matters and customer satisfaction. Moreover, IR 4.0 provides new business models by introducing Fintech facilities and digital banking.

Nonetheless, the nexus of IR 4.0 and the financial sector has created challenges regarding the regulatory authorities and framework, cybersecurity, many job abolitions, and data hacking. In this regard, a prompt and timely effort is required to tackle these important issues generated by IR 4.0.

This chapter intends to throw light on the opportunities and challenges raised due to the linkages of IR 4.0 in the financial sector. The following are the foremost objectives of this chapter:

- To examine the impact of IR 4.0 on the financial sector, including its effects on financial stability, regulatory frameworks, and societal well-being.
- To analyze the opportunities and challenges arising from the integration of IR 4.0 and finance.

DOI: 10.1201/9781003611905-5

- To identify the key drivers and barriers to the adoption of IR 4.0 technologies in the financial sector.

This chapter contributes to the existing literature on IR 4.0 and finance by providing a comprehensive and multifaceted analysis of the integration of these two fields. This chapter provokes a debate in the field of finance, technology, economics and sociology, which has become a hot topic due to the integration of IR 4.0 and finance, including the opportunities and challenges that occur because of their integration.

This integration has produced vital policy implications regarding the rules and regulations, principles and standards to develop and use IR 4.0 technologies, especially its training and workshops. The threats like cybersecurity, data breaches, abolition of jobs, and social imbalances raised from this integration should also be addressed. The essential part of this chapter is to highlight the opportunities and challenges faced by financial institutions because of the adoption of AI, blockchain, big data analytics, and IoT which have effects on the efficiency of financial institutions regarding their operations, financial inclusion, managing risk, and customer satisfaction. Through case studies and real-world applications, we illustrate how leading financial institutions are navigating this technological landscape, providing strategic recommendations for successful adaptation to IR 4.0.

3.2 BACKGROUND: IR 4.0 AND THE FINANCIAL SECTOR

The successive industrial revolutions have not only transformed the entire world but also introduced radical changes in the banking sector all over the globe. The banking system has passed through the basic structural reforms in the beginning of the First Industrial Revolution, which ultimately laid down the foundation of the modern banking system. By the end of the 19th century, the second Industrial Revolution promoted the banking sector by introducing mass manufacturing, wireless and improved transmission of communication. The third industrial revolution also known as "Digital Revolution" focused on digitalization and transformation in the banking system. The fourth industrial revolution, IR 4.0, is considered the extension of the previous one and focuses on the integration of digital financial products and financial inclusion.

The Fourth Industrial Revolution (IR 4.0) has opened the door of technological transformation, especially in the financial sector (Brynjolfsson & McAfee, 2014). IR 4.0 has redefined the financial sector particularly the banking sector (Gomber et al., 2017). The step toward the transformation brings not only the opportunities for financial institutions but also, they must face complex challenges. The financial sector has observed a storm of disruptions and advancements after the evolution of these emerging technologies all over the world. IR 4.0 has broadened the financial sector to achieve and access better financial services, customer requirements and ways of getting high revenues (KPMG, 2025). Nevertheless, the world has also faced several challenges because of these technologies such as understanding the new techniques to operate them, requirement of heavy investment to develop digital infrastructure and regulatory mechanism, etc.

The financial institutions failing to adopt IR 4.0 left behind in the financial sector. That's why, financial institutions have never before passed through such a challenging landscape. The adoption of IR 4.0 helps to explore the matchless revenues in the financial institutions and such financial institutions have made rapid growth. Moreover, the financial institutions that have blockchain, AI, and IoT are considered a benchmark in the industry.

3.3 EMERGING TECHNOLOGIES IN IR 4.0

IR 4.0 has introduced several technologies which have entirely transformed financial institutions, through which financial services, customer services, and businesses have improved. The IR 4.0 emerging technologies are as follows:

3.3.1 Artificial Intelligence (AI) and Machine Learning (ML)

Financial institutions have been facing problems like fraud, portfolio optimization, and inconsistent customer services. AI has been used to detect fraudulent activities, efficient portfolio investments, and improved customer satisfaction. Machine learning (ML) and AI are the major emerging technologies prominently contributed to the financial sector. Potential applications of AI are:

- **Fraud Detection and Risk Management:** Unexpected and unusual patterns and frauds are detected through AI and minimize the risk by using different AI algorithms. For example, Wells Fargo connected all its operations with AI to help with the various activities like aligning the call-center operations and to improve security.
- **Customer Service:** AI provides help in addressing complicated customer queries through its chatbots and digital assistants. AI enables customer support without any physical presence of human being. In this regard, an Australian-based bank has overcome 40% wait time and improved the customer services by implementing AI.
- **Investment Strategies:** Advanced AI tools analyze big datasets, up-downs of the market and offer fruitful investment decisions. In this regard, DeepSeek, a famous AI tool, is being used for making investment decisions.

3.3.2 Blockchain

Blockchain is another renowned emerging technology which offers transparent mechanism for financial transactions by using distributed ledger technology (Nakamoto, 2008). Blockchain technology is being used in different areas of finance like:

- **Cryptocurrencies:** Cryptocurrencies are being used widely. Cryptocurrencies like bitcoins are traded through different forums without any intermediaries. Blockchain-based platforms enable direct, peer-to-peer execution of financial transactions without the need for intermediaries. Such digital currencies have challenged the running banking system.

- **Smart Contracts:** Blockchains provide smart contracts whose terms and conditions are based on coding, verifying, and implementing the agreement in the absence of any agent. Such contracts reduce the transaction cost and improve efficiency.
- **Cross-Border Payments:** Blockchain also facilitates prompt and quick cross-border payments by reducing the transaction fees.

3.3.3 Big Data Analytics

The financial institutions not only collect data from customers but also gather it from all the stakeholders like financial markets, money lenders, currency exchanges, government authorities, etc. daily and they extract meaningful information through big data analytics. Basically, big data analytics facilitates accessing and analyzing the data to get important insights:

- **Customer Behavior Analysis:** The financial institutions judge the behavior of the customer by its transaction history and interaction. Big data analytics thoroughly analyzes the customer's behavior regarding its expectations, loyalty, and satisfaction.
- **Risk Assessment:** Big data analytics enables the financial institutions to assess risk before the time by analyzing the ups and downs in the market.
- **Regulatory Compliance:** In the fraudulent activities, big data analytics helps to make sure 100% regulatory compliance by monitoring and analyzing the transactions.

3.3.4 Internet of Things (IoT) and Edge Computing

Atzori et al. (2010) are in the view that IOT depends on a sensor-based software through which different devices and tools are linked to capture and transmit the data. Edge computing supports financial institutions to process the data for bringing it closer to its source. IoTs have been used in science and technology, but the financial institutions are still deprived of such advanced networks. The utilization of IoT has brought revolutionary and exemplary advancements in the field of finance. Financial institutions have still room to enhance their revenues such as:

Usage-Based Insurance: From the perspective of IoT technology, it monitors the behavior which helps the insurers regarding premiums through real-time data.

Asset Management: IoT sensors are also used to monitor the flow of assets value. It helps to make investment decisions for investors by real time following its volatility linked to underlying and external factors.

3.3.5 Cloud Computing and Cybersecurity

Cloud computing provides computing services over the internet which reduces the cost of financial institutions (Armbrust et al., 2010). On the other hand, cybersecurity provides data integrity, security, and privacy. These emerging technologies have offered several opportunities to financial institutions through innovation. But financial institutions also need to navigate the new challenges:

- **Data Storage and Accessibility:** Data storage and accessibility enable the financial institutions more information for the customers. In this regard, cloud computing provides such services to store heavy data and ensures its accessibility.
- **Security Measures:** Cyber crimes and data breaches have shook the customer confidence on the financial institutions. Cybersecurity protocols save these security threats and maintain the trust of the customer.

3.3.6 Robotic Process Automation (RPA)

It is a robotic process that automates repetitive and rules-based activities in the financial world.

- **Transaction Processing:** Generally, financial institutions must perform the same activities repeatedly over the period. The financial institutions may handle this challenge by using RPA.
- **Compliance Reporting:** Prompt and accurate compliance of regulations are required where RPA may perform it efficiently.

3.3.7 Digital Transformation

Digital transformation led by IR 4.0 provides financial institutions with more improved customer experience along with reduced costs. Financial institutions can transform data by using data analytics.

- **Credit Management:** Credit management is a vital function of financial institutions where businesses must face complicated and technical issues. These technologies assist investors to manage their debt through different digital applications.
- **Marketing Analytics:** Nowadays businesses heavily depend on marketing where customer requirements and expectations matter a lot. In this regard, these technologies enable them to sort out the data digitally. In fact, customer loyalty and brand loyalty are influenced through these technologies.

3.3.8 Fintech Partnerships

In this competitive environment, Fintech companies are making partnerships with financial institutions. Financial institutions meet their competitive needs through the expertise of Fintech partnerships.

- **Improved Risk Management and Compliance:** AI enables financial institutions to monitor their real-time transactions, fraud, scams, and their compliance.

- **Financial Inclusion:** Digital technologies enable the financial institutions to provide their services in remote areas where the physical presence of any bank may enhance the cost of the bank. In this regard, mobile applications assist to the customers in financial services.

3.3.9 Innovation Labs

Financial institutions are investing in innovative labs to observe the efficiency of these technologies. These incubator centers provide new Fintech innovations, optimization and new business ideas to improve financial services (PwC, 2019).

- **Kinexys:** Financial institutions face a challenge to move the assets and money all over the world. Kinexys provides this facility to the financial institutions.
- **Citi Innovation Lab:** These labs emphasize optimal decision-making through blockchain technology. It enables the institutional investors to explore real-time data platforms for investment purposes.
- **Wells Fargo Innovation Lab:** It is a US-based well-renowned and specialized lab in cybersecurity to detect frauds and scams in the world.

3.4 INDUSTRY 4.0: REVOLUTIONIZING THE FINANCIAL ECOSYSTEM

The beginning of IR 4.0 has nourished technological advancement, unexpected and unexplored opportunities for the financial institutions in shifting their operations and services. This revolutionary phase has integrated AI, blockchain, IoT, and quantum.

3.4.1 Enhanced Risk Management

IR 4.0 assists the financial institutions through data analytics and ML to realize and reduce the risk efficiently. Proactive measures through AI system can be used to find out the credit defaults to avoid heavy losses.

3.4.2 Personalized Financial Services

Personalized financial services to the individual customers are provided by AI and ML algorithms. Chatbots are the perfect example of it where customers can have customized investment advice.

3.4.3 Blockchain-Based Secure Transactions

Blockchain technology also makes financial transactions more secure and transparent to reduce the risk level. Payment system and international transactions can be made by developing the mechanism through blockchain technology.

3.4.4 Quantum Computing: Optimizing Portfolio Management

Portfolio management requires scaling large amounts of financial data and their probable combinations. In this context, quantum computing can offer a potential breakthrough in searching for an optimal portfolio and mitigate potential losses. The enhanced computational efficiency and problem-solving capabilities give quantum computing edge in data intensive domains like optimizing portfolios.

3.4.5 IoT-Enabled Payments

The IoT enables financial establishments to foster more innovative payment solutions, such as contactless payments and wearable devices. IoT-enabled devices can replace NFC with more advanced IOT-based tool.

Concluding the arguments, we can safely assume that IR 4.0 presents a plethora of opportunities for financial institutions to transform their operations, products, and services. By leveraging AI, quantum computing, blockchain and IoTs, financial institutions and markets can brace themselves for the more efficient risk management, personalized financial services, ensure secure transactions, optimize portfolio management, and develop innovative payment solutions.

3.4.6 Investment Strategies

IR 4.0 transforms investment strategies by fostering real-time decision-making and leveraging dynamic form of data analytics, AI, and ML. Quantitative investment approaches, such as algorithmic trading and factor-based investing, become increasingly prevalent. AI-powered systems using quantum or computational computing prowess to digest the more complex financial data to understand any hidden or obvious patterns and then predict the market flow, mood, and sentiment. Furthermore, IR 4.0 enables the development of novel investment strategies, such as IoT-generated or other non-traditional data. It also enables the creation of optimal portfolios by using algorithms and big data analytics. Quantum computing, a fundamental component of IR 4.0, further ease the formation of portfolios with balanced or most optimal risk to return ratio.

3.5 TRANSFORMATION OF EMERGING FINANCIAL SECURITIES AND IR 4.0

3.5.1 Transformation of Financial Derivatives

IR 4.0 transforms the financial derivatives terrain by introducing novel, technology-driven mechanisms. With the advent of more powerful AI, the algorithms induced by the complex computational power can help create complex derivatives, such as AI-powered options and futures contracts. Additionally, blockchain technology, introducing smart contracts that automate execution based on predefined conditions, facilitates the development of decentralized, transparent, and secure derivatives markets. This makeover enables stakeholders to handle risk associated with these securities more effectively. It also allows the investors to identify new opportunities, and access previously inaccessible markets. This substantially influences the derivatives market, transforming the way CDOs, options, futures, and swaps are engineered, negotiated, and traded.

3.5.2 Collateralized Debt Obligations (CDOs)

CDOs can redefine their markets using some of the latest developments of IR 4.0 that can enhance their valuation, risk management, and liquidity. Latest industrial trends enhance CDOs by

- **AI-driven Credit Scoring:** The transformation of traditional frameworks like Altman Z score and others into more robust and tech-based scoring models with lower transaction costs and reduced settlement times.
- **Blockchain-based Transparency:** Distributed ledger technology ensures transparent and secure ownership tracking, reducing counterparty risk.
- **Quantum Computing:** Optimizes CDO portfolio composition, reducing risk and increasing returns.

3.5.3 Options and Future Trades

IR 4.0 influences options and future trading by enabling the simulation of complex options pricing models, allowing for more accurate risk management. Similarly, new Fintech solutions can structure options pricing and predict future price movements, reducing volatility. Blockchain technology optimizes margin calls, reducing counterparty risk.

3.5.4 Hedging Industry

IR 4.0 transforms the hedging industry in several ways. Use of additional computing power with more advanced and trained algorithms allows financial commodities to find the solutions that counter the contemporary risk in other commodities. These innovative financial solutions, for example, smart contracts enable deeper insight into the volatility, systematic risk, and liquidity pattern which can help balance the overall risk of the portfolio.

3.6 IR 4.0 IN ISLAMIC FINANCE

IR 4.0 is poised to transform the Islamic finance industry by providing innovative, Shariah-compliant financial solutions. Blockchain technology, for instance, enables the creation of decentralized, transparent, and secure Islamic financial markets. AI-powered systems facilitate the development of sophisticated risk management strategies, compliant with Islamic finance principles. Furthermore, IR 4.0 enables the creation of novel, Shariah-compliant financial instruments, such as AI-powered sukuk (Islamic bonds) and Islamic derivatives.

3.6.1 Transformation of Traditional Financial Models

IR 4.0 is transforming the investment landscape by introducing new models and techniques that enhance investment portfolio management and asset valuation. The traditional Black-Scholes model, for instance, is being reevaluated in light of IR 4.0 advancements. New models, such as the Heston model and the SABR model, are being developed to provide more accurate option pricing and risk management.

The Capital Asset Pricing Model (CAPM) is also being revised to incorporate new factors and risks associated with IR 4.0. The Fama-French model, for example, adds size and value factors to the traditional CAPM, providing a more comprehensive framework for asset pricing. Additionally, new models, such as the Carhart model and the Pastor-Stambaugh model, are being developed to incorporate momentum and liquidity factors into the asset-pricing framework.

IR 4.0 is also driving the development of alternative risk management models. The Value-at-Risk (VaR) model, for instance, is being supplemented with more advanced models, such as the Expected Shortfall (ES) model and the Conditional Value-at-Risk (CVaR) model. These models provide a more comprehensive framework for risk management, enabling investors to better navigate the complexities of the IR 4.0 landscape.

Furthermore, IR 4.0 enables the development of more sophisticated factor models, which can be used to identify and manage risk factors associated with specific asset classes. The Barra model, for example, is a widely used factor model that provides a comprehensive framework for managing risk in equity portfolios. New models, such as the Axioma model and the MSCI model, are being developed to incorporate new factors and risks associated with IR 4.0.

In conclusion, IR 4.0 is transforming the investment landscape by introducing new models and techniques that enhance investment portfolio management and asset valuation. As the IR 4.0 landscape continues to evolve, investors will need to adapt and innovate to remain competitive.

3.7 BANKING INDUSTRY AND IR 4.0

3.7.1 Industry 4.0 Applications in Banking

- AI-powered fraud detection can save banks an estimated $12 billion annually in fraud prevention. Application programming interfaces of third-party financial services providers requesting the banking data is estimated to surpass 720 billion globally by 2029 (*Juniper Research, 2024*).
- The **Fintech industry** is projected to reach $324 billion by 2026, creating new economic opportunities (*CB Insights, 2023*).

Digital-Only Banking Products

- **Mobile-only Bank Accounts:** All banking services are accessible through mobile apps, eliminating the need for physical branches.
- **Digital Wallets:** Integrated wallets that enable seamless transactions, payments, and money transfers.

AI-Powered Investment Products

- **Robo-advisory Services:** Transformed the investment landscape by offering professional portfolio management services with minimum human input. Their ability to process large-scale data and use of predictive modeling make them the highlight of AI-driven financial innovation.
- **AI-powered Trading Platforms:**
- By leveraging sophisticated algorithms, AI-enabled trading platforms train on, process, and predict patterns in large datasets, ranging from market forces and macroeconomic variables to sentiment metrics, and output this analysis as charts, trend lines, and key risk-adjusted ratios.

Blockchain-Based Products

- **Blockchain-based Payment Systems:** Secure, transparent, and efficient payment systems that utilize blockchain technology.
- **Smart Contract-based Lending:** Automated lending platforms that use smart contracts to facilitate lending and borrowing.

IoT Banking Products

- **IoT-enabled Payment Devices:** Wearable devices, such as smartwatches, that enable contactless payments.
- **Smart Home Banking:** Integrated banking services that enable users to manage their finances and make payments through smart home devices.

Quantum Computing-enabled Products

- **Quantum-secured Transactions:** Ultra-secure transaction platforms that utilize quantum computing to prevent cyberattacks.
- **Quantum-optimized Portfolio Management:** Advanced portfolio management systems that use quantum computing to optimize investment strategies.

Personalized Banking Products

- **Personalized Banking Apps:** AI-powered banking apps that provide personalized financial recommendations and services.
- **Customized Credit Products:** AI-driven credit products that offer personalized credit limits, interest rates, and repayment terms.

These are just a few examples of the innovative banking products that could emerge as a result of IR 4.0. As technology continues to evolve, we can expect to see even more innovative and customer-centric banking products.

3.8 INDUSTRY 4.0 IN FINANCE AND ITS ROLE IN ACHIEVING SDGs

IR 4.0 in financial institutions aligns with multiple United Nations Sustainable Development Goals (SDGs) by fostering innovation, economic growth, financial inclusion, and responsible consumption of resources. Below is a structured analysis of how the key themes of this chapter relate to certain SDGs:

3.8.1 Financial Inclusion and SDG 1 (No Poverty) and SDG 10 (Reduced Inequalities)

How It Relates:

- Digital banking, blockchain, and AI-driven financial services enable greater access to financial services for underserved communities, reducing poverty and income inequality.
- **Example:** The Global Findex database 2021 shows that 76% of adults globally now have bank accounts, compared to 68% in 2017, largely due to digital financial services (*World Bank, 2021*).
- **IR 4.0 Applications:** Mobile banking, AI-powered micro-lending, and blockchain-based identity verification are helping expand financial access in developing economies.

Supporting Statistics:

- Digital financial inclusion has helped 1.2 billion people gain financial access in the last decade (*World Bank, 2021*).
- The use of AI-driven credit scoring in emerging markets has improved loan accessibility for small businesses by 30% (McKinsey & Company, 2021).

3.8.2 Digital Transformation and SDG 8 (Decent Work and Economic Growth)

How It Relates:

- IR 4.0 technologies improve operational efficiency and productivity in financial institutions, contributing to economic growth.
- AI and automation help create new job roles in data science, cybersecurity, and AI ethics while reducing repetitive tasks.
- **Example:** The World Economic Forum (WEF) estimates that by 2025, 85 million jobs will be displaced by automation, but 97 million new jobs will be created (WEF, 2023).

3.8.3 Sustainable Finance and SDG 9 (Industry, Innovation, and Infrastructure) and SDG 13 (Climate Action)

How It Relates:

- **Green finance** and AI-powered Environmental, Social, and Governance (ESG) investments enable sustainable banking and responsible lending.
- **Blockchain-based carbon credit trading** enhances transparency and effectiveness in global climate action.

Examples and Case Studies:

- **HSBC and BNP Paribas** have integrated AI to assess climate risk exposure, improving sustainable investment strategies.
- **Blockchain for Carbon Tracking:** IBM's blockchain-based Green Asset Wallet tracks carbon credits, ensuring credibility in sustainable finance.

Supporting Statistics:

- The green bond market reached $1.5 trillion in 2023, with blockchain improving tracking efficiency.
- AI-powered ESG analytics tools help banks reduce carbon emissions by 20% (PwC, 2023).

3.8.4 Cybersecurity, Data Protection, and SDG 16 (Peace, Justice, and Strong Institutions)

How It Relates:

- Financial institutions implementing AI-driven fraud detection and blockchain security contribute to stronger and more resilient financial systems. For example, in 2023, AI-powered cybersecurity tools helped banks reduce cyber fraud by 40%.

Key Industry 4.0 Technologies Enhancing SDG 16:

- **Blockchain** for secure financial transactions and fraud prevention.
- **AI-powered cybersecurity** tools for real-time fraud detection and compliance.
- **J.P. Morgan's AI-based fraud detection system** has saved the bank $150 million annually in fraud prevention costs (J.P. Morgan, 2024).

3.8.5 Responsible Consumption and SDG 12 (Sustainable Consumption and Production)

How It Relates:

- Cloud computing and AI in banking reduce paper waste, energy consumption, and carbon footprints by enabling digital transactions and paperless banking.

- Example: The European Bank for Reconstruction and Development (EBRD) reports that digital banking reduces energy use in banking operations by 25% (De Haas, 2023).

Key Sustainability Trends in Financial Institutions:

- **Green Banking Initiatives:** 90% of major banks now have sustainability-focused financial products (PwC, 2023).
- **AI-powered Energy Management:** Financial institutions use AI to optimize data center energy use, reducing consumption by **30%** (*Gartner, 2023*).

IR 4.0 in financial institutions aligns closely with several SDGs, particularly in the areas of financial inclusion (SDGs 1, 10), economic growth (SDG 8), sustainability (SDGs 9, 12, 13), and cybersecurity (SDG 16). As financial institutions continue adopting advanced technologies, they must integrate responsible AI, blockchain security, and green finance to maximize their contribution to global sustainability and economic development.

By leveraging AI, blockchain, big data, and automation responsibly, financial institutions can drive economic development, boost security, and adjust disparities, playing a crucial role in achieving the United Nations 2030 Agenda for SDGs.

3.9 GOVERNANCE, REGULATIONS, AND IR 4.0

The digital transformation in the recent times has led to the different challenges. Similarly advancement in industrial sector not only transformed the financial securities and markets but also the way these securities are administered. The changing landscape has significantly impacted on the primal concepts of CSR and corporate governance. Regulatory bodies, industry-led initiatives, and financial ecosystem work together to promote best practices, standards, and investor protection in the IR 4.0 era.

3.9.1 Industry-Led Initiatives

Several industry-led initiatives are launched to promote best practices and standards for IR 4.0 in the finance world, including:

1. **Financial Information Forum (FIF):** An industry-led initiative that promotes best practices and standards for financial data management.
2. **Global Financial Markets Association (GFMA):** An industry-led initiative that promotes best practices and standards for global financial markets.
3. **Institute of International Finance (IIF):** An industry-led initiative that promotes best practices and standards for global financial stability.

3.9.2 New Rules and Regulations

To govern IR 4.0 in the finance world, new rules and regulations are introduced, including:

1. **General Data Protection Regulation (GDPR):** Enhances data protection and privacy for individuals in the European Union.

2. **Payment Services Directive 2 (PSD2):** Regulates payment services and enhances security measures for online payments in the European Union.
3. **Markets in Financial Instruments Directive II (MiFID II):** Enhances transparency, accountability, and security in financial markets in the European Union.
4. **Bank Secrecy Act (BSA) and Anti-Money Laundering (AML) Regulations:** Enhanced regulations to prevent money laundering and terrorist financing in the financial sector.
5. **Cybersecurity and Infrastructure Security Agency (CISA) Regulations:** Regulations to enhance cybersecurity and protect critical infrastructure in the financial sector.

3.9.3 Advancements in Cybersecurity

Use of the digital ledgers, smart contracts, peer-to-peer encryption has greatly enhanced the cybersecurity of financial transactions and derivatives. The ethical side or the misuse of this secure network still poses challenges. New cryptographic techniques, such as lattice-based cryptography and hash-based signatures, are developed to resist quantum computing attacks. Firewalls and blockchain-based security measures have long been touted as a future to digitally secure financial instruments and markets.

3.10 CHALLENGES OF INTEGRATING IR 4.0 IN FINANCIAL INSTITUTIONS

The rise of the IR 4.0 has sparked a wave of excitement across industries. While IR 4.0 pose many prospects, their integration into financial instruments and markets presents several challenges. Financial institutions, often burdened by legacy systems and rigid structures, are finding it hard to keep pace with the sweeping technological changes. For example, technological upgrades often encounter substantial cost overruns and delays due to inadequate planning and execution.

For successful integration, a robust infrastructure and skilled labor is required. This will necessitate the transformation of the education and training which institutes are imparting. Emerging technologies like data analytics is the way forward. Developing economies struggle in this domain. They don't have the infrastructure to support the requirements of IR 4.0. These fragile economies coupled with weak and volatile financial markets increases exposure to cyber threats which can result in proliferation of AI-generated disinformation.

To keep up with the pace of technological advancements, financial institutions need to up their game of governance and accounting for unknown threats arising from techno wave. For this reason, they must navigate complex regulations that may not yet account for new technologies, requiring continuous dialogue with regulators to ensure compliance while fostering innovation.

On one side, this increases the financial inclusion and offer new products that have the potential to reach millions of consumers but on the other hand inadequate digital infrastructure, digital literacy gaps, and the gender divide can hinder their effectiveness, particularly in developing countries. The financial world needs to have this dialogue in order to maximize the benefits on offer from the IR 4.0.

The deployment of AI and other advanced technologies raises ethical concerns, including data privacy, algorithmic bias, and the potential for job displacement. Financial institutions must address these issues to maintain public trust and ensure that technological advancements do not exacerbate social inequalities.

3.10.1 Navigating the Paradigm Shift: Strategic Imperatives for Financial Institutions in Industry 4.0

The advent of IR 4.0 has precipitated a seismic shift in the financial services landscape, necessitating a fundamental transformation in the strategic orientation of financial institutions. As the velocity and complexity of technological innovation continue to accelerate, financial institutions must navigate an increasingly complex regulatory environment, mitigate burgeoning cybersecurity threats, and cultivate innovative business models that leverage the transformative potential of emerging technologies.

3.10.2 Reconfiguring Business Models: The Imperative of Digital Transformation

To remain competitively viable, financial institutions must undergo a profound digital transformation, reconfiguring their business models to accommodate the paradigmatic shifts occasioned by IR 4.0. This necessitates the deployment of cutting-edge technologies, including AI, blockchain, and the IoT, to create novel revenue streams, optimize operational efficiency, and enhance customer experience. Moreover, financial institutions must cultivate a culture of innovation, fostering a nimble and adaptive organizational mindset that facilitates the rapid iteration and deployment of new technologies and business models.

3.10.3 Mitigating Systemic Risk: Cybersecurity and Regulatory Compliance

The increasing reliance on digital technologies has created new vulnerabilities, rendering financial institutions susceptible to burgeoning cybersecurity threats. To mitigate these risks, financial institutions must implement robust cybersecurity protocols, leveraging advanced threat detection systems, encryption technologies, and incident response frameworks. Furthermore, financial institutions must navigate an increasingly complex regulatory environment, ensuring compliance with evolving industry standards and regulatory requirements. This necessitates the development of sophisticated compliance management systems, as well as ongoing engagement with regulatory bodies and industry associations to shape policy and best practices.

3.10.4 Cultivating Strategic Agility: Talent Acquisition and Development

To navigate the complexities of IR 4.0, financial institutions must cultivate strategic agility, leveraging human capital to drive innovation and growth. This necessitates the acquisition and development of top talent, possessing expertise in emerging technologies, data analytics, and digital transformation. Financial institutions must also foster a culture of continuous learning, providing ongoing training and development programs that enhance digital literacy and promote innovation. By cultivating a nimble and adaptive workforce, financial institutions can respond effectively to the rapidly evolving demands of IR 4.0.

3.10.5 Reimagining Risk Management: Emerging Technologies and Advanced Analytics

The increasing complexity of financial systems, coupled with the burgeoning risks associated with emerging technologies, necessitates a fundamental reimagining of risk management frameworks. Financial institutions must leverage advanced analytics, ML, and emerging technologies to develop more sophisticated risk models, capable of identifying and mitigating potential threats. Furthermore, financial institutions must establish robust governance and oversight frameworks, ensuring that risk management protocols are integrated into all aspects of organizational decision-making. By reimagining risk management, financial institutions can proactively mitigate the risks associated with IR 4.0, while capitalizing on its transformative potential.

3.11 FUTURE PROSPECTS, IR 5.0, AND FINANCE

The transformative shift to Industry 5.0 (IR 5.0) in the financial world is expected to integrate human centric intelligence and innovation with smart tech. IR 5.0, which focuses on the convergence of humans, machines, and AI, will enable the creation of more sophisticated financial models, risk management systems, and investment strategies.

One of the key areas where IR 5.0 will have a significant impact is in the development of more advanced AI-powered financial advisory systems. These systems will be able to provide personalized investment advice, risk management, and portfolio optimization. Additionally, IR 5.0 will enable the creation of more sophisticated chatbots and virtual assistants, which will be able to provide customers with instant support and guidance on financial matters.

Another area where IR 5.0 will have a significant impact is in the expansion of the more secure and far-reaching micro finance ecosystem. The use of advanced technologies such as blockchain, quantum computing, and the IoT will enable the creation of more secure, transparent, and efficient payment systems. These systems will be able to facilitate instant, cross-border transactions, reducing the need for intermediaries and enabling greater financial inclusion.

The integration of IR 5.0 and the financial world will also enable the creation of more sophisticated risk management systems. Advanced AI-powered systems will be able to generate many more leads that can formulate the relation between the current market standings with the upcoming challenges and potential biases in investors. Smart tech with collaborative intelligence will be the way forward in understanding the complexities of financial markets.

Furthermore, IR 5.0 will enable the creation of more personalized and customer-centric financial services. Advanced intelligent systems will be able to explore user data, behavior, and preferences, enabling financial institutions to provide more tailored services as per the changed dynamics. This will enhance consumer experience resulting in loyalty, and retention.

In short, IR 5.0 brings transformation in terms of renewed focus on human synchronization with the modern intelligence tools and automation (by products of IR 4.0). Digital economy will revolve around Web 3.0. The proxy channels or intermediaries will continue to be cannibalized by the advancement of decentralized finance (DeFi). Similarly, IR 5.0 will interlink ethical AI and hybrid models of governance and decentralized platforms.

3.11.1 How the Human Capital Will Be Impacted by the Industrial Revolution

The integration of IR 4.0 and the financial world will undoubtedly lead to significant changes in the way financial services are delivered. While IR 4.0 will automate many tasks, augment human capabilities, and increase efficiency, it is unlikely to eradicate the need for human capital. Following are some of the reasons why human capital will still be a force to reckon with:

1. **Complex Decision-making:** While AI and ML algorithms can analyze vast amounts of data, human judgment and expertise are still necessary for complex decision-making, particularly in situations involving ambiguity, uncertainty, or ethical considerations.
2. **Relationship-building and Trust:** Human relationships and trust are critical in the financial sector, particularly in areas such as wealth management, investment banking, and corporate finance. Human capital is necessary for building and maintaining these relationships.
3. **Creativity and Innovation:** Human capital is essential for driving creativity and innovation, with a vision for the future direction, in the financial sector. Professionals with expertise in finance, economics, and technology will be needed to develop new financial products, services, and business models.
4. **Regulatory and Compliance Expertise:** As industrial revolution continues to evolve, regulatory and compliance requirements will become increasingly complex. Human capital with expertise in regulatory affairs, compliance, and risk management will be essential for navigating these complexities.

3.11.2 New Roles and Skills

While IR 4.0 automates many tasks and augments human capabilities, human capital will still be essential for driving creativity, innovation, and complex decision-making in the financial sector. New roles and skills are emerging, and professionals are needed to adapt to work effectively alongside IR 4.0 technologies.

1. **Digital Transformation Specialists:** Experts with knowledge of digital transformation, change management, and organizational development will be necessary to guide financial institutions through the IR 4.0 transition. New horizons and opportunities for professionals with expertise in data science, ML, and data analytics.
2. **Cybersecurity Experts:** Professionals with expertise in cybersecurity, risk management, and compliance will be essential for protecting financial institutions and their customers from cyber threats.
3. **Human-centered Design Specialists:** Professionals with expertise in human-centered design, user experience, and behavioral economics will be necessary to develop IR 4.0 solutions that meet the needs of customers and employees.

3.12 CONCLUDING REMARKS

The confluence of IR 4.0 and finance has heralded a paradigm shift in the financial sector, reshaping its fundamental dynamics and propelling it toward an era of unprecedented technological sophistication. This chapter has delved into the transformative impact of emerging technologies—AI, blockchain, big data analytics, the IoT, and quantum computing—on the financial ecosystem. These innovations are not only enhancing operational efficiency and risk management but also revolutionizing investment strategies, regulatory compliance, and the very architecture of financial markets. However, with these advancements come formidable challenges, including cybersecurity threats, ethical dilemmas, workforce displacement, and regulatory complexities. The future of finance in the era of IR 4.0, therefore, hinges on the delicate balance between technological progress and sustainable governance frameworks.

AI and ML have emerged as cornerstones of financial transformation, enabling predictive analytics, algorithmic trading, and enhanced fraud detection mechanisms. The ability of AI-driven models to analyze vast datasets in real time has redefined portfolio management, credit assessment, and financial forecasting, significantly reducing inefficiencies and information asymmetries. However, concerns surrounding algorithmic bias, transparency, and the potential for systemic risk necessitate robust regulatory oversight and ethical AI frameworks. The integration of blockchain technology has further amplified this transformation, fostering DeFi, enhancing transactional transparency, and mitigating counterparty risks. The proliferation of smart contracts and tokenized assets is redefining financial instruments, disrupting traditional banking paradigms, and challenging existing legal and compliance frameworks.

Big data analytics, coupled with IoT, has unlocked unprecedented levels of financial personalization, enabling hyper-targeted services, real-time risk assessment, and predictive financial modeling. The synergy between these technologies has facilitated the development of real-time credit scoring mechanisms and enhanced due diligence processes, thereby broadening access to financial services and fostering greater financial inclusion. Nevertheless, the exponential growth in data generation raises pertinent concerns regarding data privacy, cybersecurity vulnerabilities, and regulatory compliance, necessitating the development of comprehensive data governance frameworks.

Quantum computing, still in its nascent stages, is poised to disrupt financial modeling, cryptographic security, and high-frequency trading. The immense computational power of quantum algorithms has the potential to optimize portfolio management, enhance Monte Carlo simulations, and revolutionize risk modeling. However, its application in cryptographic decryption presents significant cybersecurity threats, necessitating the development of quantum-resistant encryption protocols to safeguard financial systems.

Despite these technological advancements, the integration of IR 4.0 into finance is not without its challenges. Regulatory bodies worldwide are grappling with the complexities of governing these technologies, striving to strike a balance between fostering innovation and ensuring financial stability. The advent of AI-driven financial decisions, DeFi ecosystems, and automated trading systems necessitates a reevaluation of existing regulatory paradigms to mitigate systemic risks and uphold market integrity. Furthermore, the displacement of traditional financial roles due to automation underscores the imperative for workforce reskilling, emphasizing digital literacy, cybersecurity proficiency, and AI governance expertise.

Looking ahead, the transition from IR 4.0 to IR 5.0 will further redefine the financial sector by emphasizing human-AI collaboration, sustainable finance, and ethical AI deployment. The financial institutions that successfully navigate this transition will be those that embrace continuous innovation, integrate responsible AI practices, and establish resilient governance frameworks. The trajectory of financial evolution in the IR 4.0 era is unequivocally clear: the institutions that harness technological advancements while maintaining ethical and regulatory integrity will emerge as the vanguards of the future financial ecosystem.

REFERENCES

Armbrust, M., Fox, A., Griffith, R., Joseph, A. D., Katz, R., Konwinski, A., … & Zaharia, M. (2010). "A view of cloud computing." *Communications of the ACM*, 53(4), 50–58.

Atzori, L., Iera, A., & Morabito, G. (2010). "The internet of things: A survey." *Computer Networks*, 54(15), 2787–2805.

Brynjolfsson, E., & McAfee, A. (2014). *The second machine age: Work, progress, and prosperity in a time of brilliant technologies*. W.W. Norton & Company.

CB Insights. (2023). *Artificial intelligence report: Q1 2023*. https://s3-us-west-2.amazonaws.com/cbi-content/reports/CB-Insights_Artificial-Intelligence-Report-Q1-2023.pdf

De Haas, R. (2023). *Sustainable banking*. European Bank for Reconstruction and Development (EBRD).

Gartner. (2023). *The current state of AI use within finance: 2023 insights*. Gartner. https://www.gartner.com/en/documents/4803731

Gomber, P., Koch, J. A., & Siering, M. (2017). "Digital finance and FinTech: Current research and future research directions." *Journal of Business Economics*, 87(5), 537–580.

Hasan, M. Z., Hussain, M. Z., Umair, S., & Waqas, U. (2024). Understanding artificial intelligence in supply chain and innovation performance. In *Human perspectives of Industry 4.0 organizations: Reviewing sustainable performance*. https://doi.org/10.1201/9781032616810-7

J.P. Morgan. (2024). *AI boosting payments efficiency & cutting fraud*. J.P. Morgan. https://www.jpmorgan.com/insights/payments/payments-optimization/ai-payments-efficiency-fraud-reduction

Juniper Research. (2024). *Open banking: API market trends, key opportunities & forecasts 2024-2028.* https://www.juniperresearch.com/research/fintech-payments/banking/open-banking-apis-market-research-report/

KPMG. (2025). *Pulse of Fintech H2'24: Global fintech investment trends.* KPMG. https://kpmg.com/xx/en/what-we-do/industries/financial-services/pulse-of-fintech.html

McKinsey & Company. (2021). *Building the AI bank of the future.* McKinsey & Company. https://www.mckinsey.com/~/media/mckinsey/industries/financial%20services/our%20insights/building%20the%20ai%20bank%20of%20the%20future/building-the-ai-bank-of-the-future.pdf

Nakamoto, S. (2008). "Bitcoin: A Peer-to-Peer Electronic Cash System." Available at: https://bitcoin.org/bitcoin.pdf

PricewaterhouseCoopers (PwC). (2019). *Global fintech report 2019.* PwC. https://www.pwc.com/gx/en/industries/financial-services/assets/pwc-global-fintech-report-2019.pdf

PricewaterhouseCoopers (PwC). (2023). *Driving data with purpose: Why digital technology is critical to ESG success.* PwC Middle East. https://www.pwc.com/m1/en/publications/driving-data-with-purpose.html

World Bank. (2021). *The Global Findex Database 2021: Financial inclusion, digital payments, and resilience in the age of COVID-19.* https://www.worldbank.org/en/publication/globalfindex

World Economic Forum (WEF). (2023). "Future of Jobs Report 2023." https://www.weforum.org/publications/the-future-of-jobs-report-2023/

4 Building a Sustainable Future

Human Capital and Industry 4.0

Muhammad Ziaullah, Zia Ur Rehman, and Asghar Hayat

4.1 INTRODUCTION TO INDUSTRY 4.0

4.1.1 Defining Industry 4.0

Industry 4.0 represents a radical transformation of our manufacturing and industrial experience as a technological revolution (Pozzi et al., 2023). The transformation is stimulated by the convergence of critical technologies such as cyber-physical systems (CPS), the Internet of Things (IoT), big data analytics, cloud computing, and artificial intelligence (AI). There are no synonyms that could capture the essence of such a term, but it is all about the interconnectivity of these technologies for the design, optimization, and management of industrial operations (Usmani et al., 2024).

Thus, the fourth of industry created intelligent industries with advanced machines, systems, and products that communicate with one another in real time. This interconnectedness enables data-driven decision-making, enhances productivity, optimizes resource management, and facilitates predictive maintenance. Leading to the faster and more agile manufacturing processes that should rapidly adapt to market realities and customer needs (Hu et al., 2024).

Beyond operational efficiency, Industry 4.0 drives environmental, social, and economic impacts. In these aspects AI solutions allow cost reductions and higher product quality (Shabur, 2024). This paradigm changes from traditional manufacturing to green manufacturing, along with a balance in waste reduction, carbon footprint reduction, increased energy effectiveness, recycling and item tracing, has benefited the circular economy from an environmental viewpoint.

More importantly, Industry 4.0 has provided socially safe and healthy working environments for all. The employers hold up work-life balance, and the workspace safety is handled by employers (Blithe, 2023). Their working conditions have improved, their jobs are now more satisfying, and their quality of life has transformed through the use of AI, IoT, big data analytics, and cyber security. But it should not be the case that maintaining an undistracted work environment add features to normal safety protocols and makes the day-to-day industry operations hassle-free and sustainable.

DOI: 10.1201/9781003611905-6

4.1.2 The Role of Human Capital in Industry 4.0

The global social landscape and economy have been significantly impacted by the transition to Industry 4.0, a natural and expected consequence of the digital revolution in IoT, AI, big data analytics, and robotics. Although the transition poses monumental challenges, it also provides unprecedented opportunities. Beyond the economy, there is the central role of human capital in promoting environmental stewardship and social well-being, and the results come in shape of good health, education, access to quality and inclusion (Bykova et al., 2024). Therefore, in order to truly achieve the sustainable development goals (SDGs), we need to bring up the social and environmental perspectives beyond the economic metric because human needs, community well-being, and environmental care are deeply connected with each other.

4.1.3 Workforce Adaptation and Digital Competencies

The adoptability and development of digital literacy skills is shaping traditional jobs, and these are closely linked with organizational success in this fast pace of working styles (Cirillo et al., 2023). The attitudes of employees acquiring these skills depend upon different factors that influence how well an organization is coping with that. With the rise of Industry 4.0 and the shift to more flexible work models where employees can work remotely or hybrid, there is need to balance tech adoption with skill building has never been more apparent (Ahmed & Smith, 2023).

4.1.4 A Strategic Approach to Human Capital Development

To keep pace with today's fact changing digital working environment, organizations are striving to cope with the needs of this change with a well-rounded strategy to develop their people accordingly. This means putting a strong focus on lifelong learning, so that the employee's stay is adaptable and competitive. Encouraging innovation and continuous growth not only strengthens individuals' capabilities but also helps build a more resilient and productive workforce and efforts of organizations were also supporting the SDGs outlined by United Nations, reinforcing the importance of aligning business strategies with social progress including:

- **SDG 4** (Quality Education)
- **SDG 3** (Good Health and Well-being)
- **SDG 8** (Decent Work and Economic Growth)
- **SDG 10** (Reduced Inequalities)

4.2 THE DYNAMICS OF INDUSTRY 4.0

4.2.1 The Dynamics of Industry 4.0

The fourth industrial revolution, known as Industry 4.0, makes a major shift driven by the integration of advance technologies like IoT, AI, and advance automations systems. As these technologies evolve, they have not only streamlined the operations but also pushing the industries to become more productive and innovative.

This does not come to true without parallel investment in people in shape of providing digital skills, education, training, and reskilling programs to adapt the new according to the demand of the Industry 4.0. This chapter provides an in-depth analysis of these key technological drivers of Industry 4.0 like AI, IoT, Advanced Automation, etc.

4.2.1.1 AI in Industry 4.0

AI involves in systems that can mimic human abilities like reasoning, learning, and making decisions like:

- AI analyzes through real-time sensor data to spot out the potential issues and share remedies to overcome or downtime them (Rane et al., 2023). This proactive approach helps the companies to reduce maintenance costs and keep their operations to run smoothly.
- With the ability to sift through massive datasets, it is AI that can identify the inefficiencies and recommend real-time adjustments to optimize production.
- It enhances the capabilities of through robotics technology which allows them to handle more complex tasks and respond better to changing environments.
- AI tools help business to forecast demand, manage its stocks, and streamline the logistics (Verma, 2024). With predictive insights, companies can grow, remove inefficiencies, and respond quickly to disruptions and also plan effectively.

4.2.1.2 The IoT in Industry 4.0

The IoT refers a network which consists of connected devices that uses sensors and operated through different software to collect and share the data (Magara & Zhou, 2024). In context of Industry 4.0, the IoT plays a vital role by transforming how data is collected, analyzed, and acted upon. So the key applications of IoT are:

- IoT devices collect large volume of data from operational areas, production lines, equipment, etc. to help and uncover the insights about performance, quality, and efficiency of the operations which ultimately help to improve decision-making.
- With IoT, organizations can constantly track equipment and their processes. The prompt response of IoT allows to detect issues and timely interventions help to avoid costly downtime.
- It enabled the interconnected devices to better tracking of goods in transit and improve visibility and streamline the logistics.
- It also tracks the customer experiences and provides more rapid responses to boosting customer satisfaction and loyalty.

4.2.1.3 Advanced Automation in the Modern Industrial Era

Automation lies at the heart of the Industry 4.0 (Ansari et al., 2023), with advanced systems reshaping how tasks are performed across sectors, like:

- Robotics: Industrial robotics handle repetitive, physically demanding, and hazardous tasks which help to boost up the output at a lowest labor cost.

- CNC machines: These computerized machines automate precision manufacturing and increase both speed and accuracy in the production.
- Automated Guided Vehicles (AGVs) move materials and products throughout factories and warehouses without human interventions, improving logistics and cutting costs.
- Automated Storage and Retrieval Systems (AS/RS): These systems streamline how goods are stored and retrieved in warehouses, making storage more efficient and space effective.
- CPSs connect physical operations with digital systems, enabling smart control and real-time adjustments in industrial settings.

4.2.2 The Convergence of Digital Technologies and Traditional Industries

The merging of digital technology with long-established industries is transforming the global marketplace (Hrytsevich & Tahir, 2025)while this shift openshe door to new possibilities. This also brings complexed technologies likes IoT, AI, big data, and cloud computing which foster the businesses to think out how they will operate in this technological era of Industry 4.0 (Logeswaran et al., 2024). This integration is not without its demands. Companies need to build strong digital strategies, invest in latest technologies, and focus on upskilling their workforces to keep the pace. By tackling these challenges head-on and making the most of the emerging opportunities, industries can set themselves for sustainable growth, improved efficiency, and a stronger standing in an increasingly competitiveglobal economy.

4.2.2.1 The Transformative Power of Digital Technologies

Digital tools are pushing businesses operating in traditional businesses to rethink on how do they operate and do their business in this technological era. Sectors like manufacturing, retails, healthcare, and finance are seeing dramatic shifts as digital adoption accelerates.

4.2.2.1.1 Opportunities Presented by Digital Integration

- Automation, IoT, AI, and machine learning streamline business operations, cut costs, and improve output (Qudus, 2025). IoT-powered smart factories enable real-time production monitoring, reducing waste and optimizing resources.
- Digital innovation has led to new models like e-commerce, subscription services, and platform-based economies.
- Traditional businesses can tap into online marketplaces, cloud services, and digital payments to broaden their reach and grow profits (Luz, 2025).
- Technologies such as blockchain, AI, and IoT are transforming supply chains by increasing transparency boosts logistical control, while AI-powered forecasting helps reduce disruption and lower costs.

4.2.2.1.2 Challenges in Digital Transformation

- Implementing advanced technologies demand sufficient investment in terms of infrastructure, tools, and even in upskilling the workforce.
- As automation is replacing routine tasks day by day, employees need to be engaged in digital literacy initiatives to cope with this digital transition and secure job displacements.
- Greater dependence on digitalization increases the cyberattacks and data breaches. Therefore, strong protocols and regulatory compliances are vital to protect the sensitive data.
- Regulators must keep pace with technologies advancements. Effective policies needed to ensure ethical usage of IoT, AI, and support fair market practices.
- Organizational internal resistance can hinder the digital transformation. Here the role of leadership comes to play, that must cultivate a culture of innovation and encourage employees as team to embrace technological changes.

4.3 THE IMPERATIVE OF SUSTAINABILITY

4.3.1 The Need for Sustainability in Context of Industry 4.0: A Tripartite Perspective

The revolution of Industry 4.0 with its fourth wave of transformative technologies is growing rapidly, it is imperative that we should ask a critical question: Can innovation truly be considered progress if it comes at the expense of sustainability? The answer lies in our ability to align the disruptive potential of Industry 4.0 with the foundational pillars of sustainable development, for example, environmental protection, social equity, and economic resilience.

4.3.1.1 Environmental Sustainability: Balance Efficiency with Impact

The promises of Industry 4.0 increased the operational efficiency of the business and increased the smarter usage of resources (Hayat et al., 2023; Parashar et al., 2023). However, these not come without caveats. While the advanced analytical and automation have enabled real-time optimization and reduced the material wastage (Chukwunweike et al., 2024), the increased utilization of energy resources to keep the large-scale digital infrastructure (data centers, and high-frequency manufacturing systems), poses a significant threat to environment (Single et al., 2023). The proliferation of e-waste and the potential effects of ultra-customized production models are especially problematic; even in different sectors like maritime logistics, which benefit from digital efficiency, must confront the challenge of lifecycle emissions and hidden environmental costs (Ajekwene et al., 2022). Environmental sustainability is therefore cannot be afterthought as it must be embedded within the design and development of the Industry 4.0 technologies, circular economy principles and integrated assessments are the essential to avoid trading short-term gains for long-term degradation.

4.3.1.2 Social Sustainability: The Human Side of Technological Transformation

The most immediate and visible, but often overshadowed by technical and economic narratives is the social dimension of the Industry 4.0. A truly sustainable Industry 4.0 must prioritize inclusive practices of providing automation reskilling and upskilling the workers with robust opportunities by addressing the needs of vulnerable sectors and fostering cross-sector collaboration. Unfortunately, the social dimension is frequently marginalized in sustainability discourse and this omission could prove costly (Khan et al., 2023). Therefore, encouragingly, the strategic adoption of Industry 4.0 in sectors like agri-food has shown promise for enhancing HR practices, and promote equality (Mishra & Pathak, 2024).

4.3.1.3 Economic Sustainability: Navigating Growth and Risk

Increased productivity, operational efficiency, and innovative business models are the economic potentials of Industry 4.0. However, realizing the benefits requires navigating significant finance hurdles, especially for small- and medium-sized enterprises like the upfront costs of building the required infrastructure, uncertain return on investment, and the pace of technological obsolescence present persistent risks. Tailored policy interventions, targeted subsidies, and capacity building initiatives are essential to bridge up this gap (Zhao et al., 2023).

4.3.1.4 The Interconnectedness of Triple-Bottom-Line in Action

What comes to clear is that these triple-bottom-line (TBL; three dimensions) are not siloed, these are deeply interconnected from enhancing the worker's skills (social) to improve productive (economic) and to reduce operational waste (environmental) (Zhao et al., 2023). Investing in green innovation (environmental) drives cost saving (economic) and public trust (social) (Arvind Mallik, 2024). Therefore, to evaluate the Industry 4.0, this interdependency of TBL framework offers a multidimensional lens which lies at the heart of TBL.

4.3.2 Emphasizing the Role of Sustainable Practices in Ensuring Long-Term Resilience and Prosperity

In the current era of Industry 4.0, sustainability is no more an option, it becomes the necessity of the age. The change from traditional to digitalization, automation, and intelligences system has changed the societies and industries. Without the anchor of sustainable practices, this advancement in operational efficiency and the usage of technological advancements become forces to disruption rather than progress. Therefore, in order to ensure the long-term resilience and prosperity, the sustainability practices must be ensured and deeply embedded into the core of Industry 4.0 implementations.

4.3.2.1 Environmental Sustainability: Building a Resilience Ecological Foundation

Environmental sustainability always serves as a base for long-term resilience because our ecosystem already is under immense pressure of pollution, climate change, and resource scarcity. It cannot bear the unchecked burden of Industry 4.0 intensification. Because in large consumption of energy resources, e-waste and material throughout are the big draw backs which we are bearing in compensation of operational efficiency and enhanced productivity. However, keeping in loop all the technologies, these can become a powerful tool for ecological resilience like, smart farming system in agri-food sector can reduce the utilization of fertilizers and minimize the wastage of water. Therefore, the concept of circular-economy in combination with Industry 4.0 provides a clear guideline for decoupling the economic activity without harming environment (Kopeinig et al., 2024).

Green innovations lie at the heart of sustainable transformation because in Industry 4.0 technologies enable to find out the renewable energy solutions like biotechnologies and eco-friendly materials. And at the same stage, Green Marketing, not as a cosmetic exercise but also works as a strategic tool to shift the consumer behavior toward eco-friendly behavior and builds the markets around the sustainability practices (Muchenje et al., 2023). Yet, when sustainability becomes a value proposition, and this progress must remain vigilant to greenwashing risks and ensure that the marketing claims are backed by measurable outcomes (El Merroun et al., 2022; Veile et al., 2021).

4.3.2.2 Social Sustainability: The Human Factor in Future-Ready Societies

The social implications of Industry 4.0 are perhaps its most profound like from the automation and AI-driven processes are shaping labor markets and showing new jobs driven by new roles that required advance digital skills. Therefore, without social planning, this leads to inequalities and marginalized the vulnerable groups. At that stage organizations need to strategically invest in education and workforce reskilling as a paramount which is not only required to fit a person for the required job but also enable them to foster digital literacy and building a resilient workforce.

4.3.2.3 Economic Sustainability: Growth Anchored in Responsibility

Ranging from operational efficiency and enhanced production toinnovation and economci gains, these are among the key benefits of Industry 4.0. Yet, the prosperity buildup on an unsustainable foundation is inherently fragile. SMEs often constrained by limited resources, face unique challenges in adapting advance technologies. Therefore they required tailored support, accessible financing and share knowledge platforms to ensure inclusive growth and prevent the digital divide from becoming an economic chasm (Elzawi et al., 2012; Santos et al., 2023).

4.3.2.4 Interconnected Sustainability: The Triple-Bottom-Line in Action

The true resilience is emerged from TBL that are not separate ambitions but mutually reinforcing systems. The poor environmental conditions, social inequality, and unsustainable economic growth lead to undermine public health, instability, ecological and social crises, respectively. This is Triple Bottom Line (TBL) framework which provides structured way to integrate these three pillars into business and policy decision-making, aligning closely with the concept of sustainability 4.0 (Ansari; Santos et al., 2023).

4.4 HUMAN CAPITAL DEVELOPMENT IN THE DIGITAL AGE

4.4.1 Examining the Evolving Nature of Work and Skills Requirement in the Era of Industry 4.0

In today's rapidly evolving landscape, permanence is elusive. In history when once, the mastery of a trade could carry a person from apprentice to retiree, their craft as stable as the seasons. What is learned today may be absolute tomorrow. Yes, we are now living in the era of Industry 4.0, where the fusion of CPSs, the intelligence network, and AI is not just transforming the production procedures but also redefining how the work, and more critically, who we need to become the thrive. Driven by technological convergence, the fourth industrial revolution does not only replaced the muscle with machinery but also it has provided the increasing substitutes predictable cognitive work with automation, data-driven decision-making, and with the usage of AI, every corner of the global economy is being reengineered.

The result? A need for talent that understands not only how to build these systems but also have the ability to how to think with them analytically, ethically, and strategically (Shan & Wang, 2024). Engineering education is being reshaped by virtual and augmented reality, creating immersive environments that redefine experiential learning. Manufacturing now relies on digital twins, autonomous robotics, and predictive maintenance (Mayer et al., 2025; Shaikh). In logistics, maritime sectors, and even the food supply chain, technology influence is evident. Yet, the resistance does not unfold uniformly with unique skills, different sectors seek unique expertise with respect to their circular economy models. Consequently, nearly all sectors highlight the need for tailored training programs and agile education systems capable of adapting repidly to technological change.

4.4.1.1 From Hard Skills to Hybrid Talents

Coders, data scientists, cybersecurity specialists, and automation engineers are the architects of this new world and the allure of technical skills is undeniable. And Industry 4.0 taught us that technology alone does not suffice. In a world of Industry 4.0 where AI handles the "what," human workers must master the "why" and the "how." Problem solving, creative thinking, communication, and adaptability are the backbones for employability in this new era of employment where soft skills are no longer supplemental skills, but these are replaced with essential capabilities (Challoumis, 2024; Dean & East, 2019). A multidisciplinary team can tackle complex sustainability challenges, leadership, emotional intelligence, and strategic foresight become as critical as technical acumen Osuya & Cheta-Maclean, 2024).

4.4.1.2 The Human-Centered Imperative

Industry 4.0 is already forming on the horizon, promises such as recalibration. It calls for human-centric innovation, where technologies amplify human values rather than replace them. It represents the designing of work environments that are inclusive, equitable, and sustainable which refers to building ethical frameworks that foster diverse teams to lead innovations and uplift it rather than divides.

4.4.1.2.1 *Bridging the Divide: The Skills Gap and the Just Transition*

The skill gap looms large and the road ahead is too steep especially for small and medium enterprises that often lack of resources for robust digital transformation. Many of the workers are lacking behind due to non-provision of digital skills and their roles needs to be automated before the retaining programs can reach them. This divide can be bridged through proactive strategies like:

- There is need to invest in skilling and reskills through education and training, especially in underserved regions.
- There is need to design the private-public partnerships to design the curriculum which should be aligned with the needs of Industry 4.0.
- The government needs to formulate such policies that support the displaced workers with social safety nets and through retaining programs.
- There is need to change the organizational cultures to embrace the learning, experimental and adaptability culture, where failure should be treated as a part of growth cycle.

4.4.2 The Lifelong Learning—Investing in Human Capital for Sustainable Future

The investment in human capital development through education, training, and learning the emerging technologies are not a choice in this era of Industry 4.0 but these are essentials. While the industry 4.0 revolution offers significant opportunities, it also presents considerable risks. The rapid integration of intelligent systems, IoT, and the AI in the industrial operations have fundamentally changed the working environment and working styles of the workers (Van Hoang, 2024). As traditional roles are redefined, the mismatch between current competencies and future needs become start. Studies indicate that without upskilling and targeted reskilling, the skills gap is undermining productivity, competitiveness, and equitable access to opportunities (Lakshmi et al., 2024). It is automation that is streamlining the tasks, yet at the same pace it is displacing the jobs of manual labor, but it is not the end of work, yet it is a signal to reimagine it (Narkhede et al., 2024). Therefore, to address this disruption, a comprehensive, and multi-stakeholder response is essential like:

- Government needs to lead in provision of vocational training (Shaheen et al., 2023), crafting national skills strategies (Hayyat et al., 2023), and incentivizing organizations to invest in talent development.
- There is need to emerge the technologies into the pedagogical and educational techniques and promote them as cross-disciplinary skill development.
- Industry and employers must recognize the workforce development not as a cost but as a strategic investment in agility and resilience.

4.4.2.1 The Era of Lifelong Learning

In Industry 4.0, learning is a continuous process that is defining the educational and learning techniques as a dynamic lifelong learning journey. Today is the pace of micro-learning, modular programs, and digitalization where students, people, and even workers are learning in digital learning environment and it's become a new norm to use virtual and augmented reality as an immense tool of learning. These imersive technologies enable the experimental learning, significantly enhansing skill acquisition and knowledge retention. As the Industry 4.0 demands a future-ready workforce with advanced and adaptive capabilities, the access to skills alone is no longer sufficient. Therefore, this is dire neeed to cultivate a culture of continuous learning (Shaheen et al., 2023) embedded into every layer of society as well as at:

- Individual level that must take ownership of their development and growth with mindset of curiosity and adoptability.
- Organization must foster the supportive environment and develop a culture of continuous learning, reward, and encouragement into the workforce.
- Institutions must provide flexible accessible options for all to get and cater diverse learning opportunities.

4.4.2.2 Human Capital and Society Equity

Social inclusion and justice are the key factors in the development of human capital because by providing access to relevant skills and opportunities, it enables the individuals to participate and contribute the digital economy. The inclusive training programs for rural workers, women, and older employees who may otherwise be left behind are essential for ensuring that Industry 4.0 is taking all together rather than dividing (Da Roit & Iannuzzi, 2023). Studies prove that the future of sustainable work depends on the integration of human resource strategies with technological driven innovations placing human are at the heart of organizatinal strategy (Kess-Momoh et al., 2024; Ren et al., 2023; Sołtysik et al., 2024). Perhaps the most overlooked dimension of the human capital is their role in advancing the environmental sustainability. People need to be trainees in green manufacturing, resource optimization, and circular economy practices, because technology alone cannot deliver the green outcomes (Cheng et al., 2023; Li et al., 2023). There is need a for a workforce that understand that not only how to operate the advanced machinesbut also possesses the awareness and compeencies requierd tominimize the environmental impacts of its operations, ensuring a sustainble and the green future in this era of Industry 4.0.

4.4.2.3 Challenges and the Path Ahead

The past is not without obstacles, where the benefits of human capital development are clear like:

- Cost barriers continue to limit the access for small and medium enterprises.
- There are challenges for institutions to update the curriculum as there is need to keep the update in curricula with technological shifts.
- Wages disparities come to face, the people with high technical ends and those who left behind will face a big difference in wages.
- Job displacement remain a real risk, and cannot be avoided without deliberate, and inclusive policy frameworks.

4.5 LEVERAGING TECHNOLOGY FOR SUSTAINABILITY

4.5.1 Exploring How Digital Technologies can Be Harnessed to Address Sustainability Challenges (Environmental Monitoring)

In the era of Industry 4.0 revolution, technology is no longer confined to the realm of productivity and profit but it is now poised as a powerful ally of sustainability globally. From climate change mitigation to biodiversity conversation and resource optimization, these are digital technologies which are reshaping how we understand, interact, and preserve the nature (Bakker, 2024; Joseph et al., 2025). In this section, we explore how Industry 4.0 are being harnessed to address the pressing sustainability challenges and create smarter systems through AI or through IoT across sectors.

4.5.1.1 Making the Invisible Visible (Environmental Monitoring)

Environmental challenges are always complex, like pollution which is spreading silently, clime patterns which are shifting subtly, and the ecosystem with is being degraded with the passage of time; and often beyond the resolution of traditional monitoring methods (Van Doren et al., 2023). Here, digital technologies step in as a game changer. Satellite imagery, sensors, drones, and data analytics platforms have empowered the architects of this digital era to visualize and respond to environmental changes with unprecedented precision and speed. The environmental sensors are connected through IoT networks and deployed to forecast pH, soil moisture, noise pollution, and temperature.

High-resolution satellite imagery, combined with AI-driven analysis, allows the scientists for detailed mapping of land use changes, deforestation, costal erosion, and about melting glaciers. Different organizations like NASA is providing open-access satellite data to government and non-government organizations to track the wildfire, illegal mining, and crop health at scale (de Oliveira et al., 2023). Drones equipped with multi-spectral cameras are offering granular, local scale monitoring, particularly in agriculture and forestry. They detect stress in crops, and damages after storms. Together these technologies are marking the previous invisible dimensions to ecological health visible and manageable.

4.5.1.2 Resource Management (Smart Systems for Sustainable Use)

Resource depletion is one the most important global sustainability concerns driven by immense pressure of agriculture, industry, and urban development on water, energy, and raw material (Rani et al., 2023). Therefore, leveraging digital tools for efficient resource management is essential to move from traditional economy to circular economy. Cities like Log Angeles and Singapore have adopted AI-powered waste management system that monitor the usage, leakage, and demand of water (Bandara, 2023). Similarly, through implementation of smart irrigation systems, the soil moisture sensors can predict the requirement of water needs for plantation and AI-powered weather forecast can be used to deliver the water only when and where it is needed, this is reducing the consumption of waste up to 30%.

The rise of smarts grids for real-time data about electricity distribution and utilization are being used which are powered by machine learning algorithms

(Mazhar et al., 2023). These grids balance the energy loads, integrate renewable sources like wind and solar technologies and reduce the peak demand pressure of electricity. The industrial facilities use digital twins to simulate operations and identify energy inefficiencies, slashing carbon emissions without compromising performance. These are digital platforms that enable the tracking of products and materials across their lifecycle. Blockchain technology is being used to ensure transparency in supply chains, while AI predicts equipment lifespans and enables predictive maintenance; extending asset longevity and minimizing waste (Aditya et al., 2024).

4.5.1.3 Energy Efficiency: Decoding Consumption and Scaling Renewables

The energy generation systems are undergoing through a radical transformation like from traditional fossil-fuel-based energy generation models to digitally managed networks powered by clean energy (Knell, 2024). Here technologies are playing a big role in accelerating this shift and ensuring the sustainability. The big infrastructures are using IoT devices to control the lighting, heating, and cooling based on occupancy and time of day. Like in commercial buildings the use of automation is reducing the energy usage by up to 40% and significantly contributing to maintain the climate targets (Vandenbogaerde et al., 2023). At the same pace, the AI-powered platforms optimize the energy production and consumption patterns across communities, industries, and cities. The forecast of demand, monitoring of its utilization through AI-powered platform, and its supply from renewable energy resources like solar and wind sources automatically adjust the storage and making this renewable integration seamless and efficient. And consumers with the active participant in energy management techniques like usage of smart meters, and appliance can provide their feedback on energy usage. This controlled supply and usage of electricity is changing their behavior and setting the consumption patterns.

4.5.1.4 The Power of Predictive Analytics in Environmental Planning

These are digital technologies that are offering the potential to anticipate and mitigate future sustainability challenges. Through predictive analytics, now we have risk assessments, powered by big data analytics and machine learning, we can build a scenario models and make strategic planning for it (Aljohani, 2023; Emma, 2024;. There are early warning systems powered by AI analyze seismic, meteorological, hydrological data to predict earthquakes, hurricanes, and floods. Government and other stakeholders can take preventive actions to safeguard the infrastructure and lives through the best usage of these technologies (He et al., 2023).

4.6 CORPORATE RESPONSIBILITY AND ETHICAL CONSIDERATIONS

4.6.1 The Role of Corporations in Driving Sustainable Practices and Ethical Behavior in the Digital Economy

The digital economy powered by AI, blockchain, and IoT has revolutionized the global business operations and, yet it brings profound ethical and environmental challenges. Organizations are now aligning their strategies with sustainability and societal well-being along with technological development (Scheyvens et al., 2023). As the digital economy continues to expand, reshaping and accelerating innovation, corporations increasingly find themselves at a crossroads between rapid technological advancement and their ethical, social and environmental responsibilities. (Rasul, 2024). They are no longer judged solely by their profitability, but by their impact on people, the planet, and the ethical framework within which they operate. In this section, we examine how businesses can harness digital tools for sustainable growth while addressing the global and emerging ethical dilemmas.

4.6.1.1 The Dual Edge of Digital Transformation

Digital technologies like AI-driven automation not only offers unprecedented efficiency but also introduces sustainability risks like optimize usage of energy in manufacturing and reducing carbon foot prints. However, on the other side, if these are controlled to AI-powered centers, it can contribute in optimum utilization of energy resources. Similarly, while blockchain technologies enhances supply chain transparency, the proliferation of electronic devices accelerates e-waste. Therefore, corporations must adopt a dual strategy that leverage technology to advance sustainability objectives while mitigating theirassociated environmental and ecological costs (Jeffers, 2010; Ullah et al., 2024).

4.6.1.2 Ethical Imperatives in the Digital Age

- ***Algorithmic Accountability***

  AI systems can reinforce biases in hiring, lending, and law enforcement if unchecked. Companies like Microsoft now implementing Fairness Audits and diverse training datasets to ensure equitable AI outcomes (Gupta, 2024).
- ***Data Privacy and Security***

  The Cambridge Analytica scandal underscored the perils of unregulated data harvesting. Regulations like the General Data Protection Regulations (GDPR) mandate corporate accountability, but ethical responsibility extends beyond compliance. Transparency in data usage such as anonymizing customer data builds public trust (Tursunbayeva et al., 2022).

4.6.1.3 Sustainability as Strategy

Leading organization are now embedding sustainability directly into their core strategies because they are considering sustainability as a central factor for long-term competitiveness. Starting from zero carbon goals to invest in green technologies, companies are aligning their business models with the United Nations SDGs. There are the digital technologies, when used with responsibility, can significantly enhance the sustainability effects (Napoli, 2023). Organizations are using predictive analytics to optimize the supply chains to reduce the carbon omissions. The IoT sensors can easily monitor and reduce the consumption of energy resources with their real-time controls and blockchain can trace the ethical souring of materials. Yet, with these innovations there come some ethical considerations. AI systems that sort job applications, analyze the consumer behaviors, and/or automate the customer services must be designed to avoid the bias and ensure the privacy and promote the fairness.

4.6.1.4 Corporate Ethics in the Digital Age

It is Industry 4.0, that has intensified the ethical questions around data ownership, algorithmic bias, and the future of work (Chirita & Sarpe, 2024). Organizations need to go beyond the surface level of ethics codes and embed the ethical considerations into every layer of decision-making process. This means that organizations need to:

- Design AI and machine learning system that provide transparency and accountability.
- Provide respect to user data privacy and offer the in/our choices to customers, either their data be used or not. They have option to put them in or out of the data.
- Ensure that automation augments the presence of human labor rather than support the displacement of labor.
- Promote the diversity and inclusion in the organizational culture in both sectors either in technological perspective or in deployment.

4.6.2 The Role of Corporations in Driving Sustainable Practices and Ethical Behavior in the Digital Economy

The Industry 4.0 revolution is sweeping through its digital transformation. Therefore, for corporations navigating the landscape of Industry 4.0, the transparency, stakeholder's engagement, and imperative transparency no longer as an option (Wang et al., 2024) but should be taken as essential pillars of sustainability and ethical business conduct in this digital age.

4.6.2.1 Corporate Governance, Transparency, and Stakeholder Engagement

Today, corporations have access to powerful digital technologies which are capable of enhancing the sustainability efforts across every dimension of their operations from the optimizing energy consumption to enhance the supply chain transparency through digital tools like AI, IoT, and blockchains. Manufacturers are using predictive analytics to minimize the resource waste and reducing their footprints on environment (Enyejo et al., 2024). They are also using blockchains to ensure that the traceability of raw material be ensured, empowering the firms to source ethically and transparently. However, the growing dependence of digital technologies is also increasing the environmental costs. The expansion of data centers, the challenges of e-waste disposal, and the proliferation of digital devices have raised new concerns about sustainability of digital infrastructure (Ramasamy et al., 2024); organizations need to adopt holistic approaches to minimize their impact through thoughtful designs, responsible procurement, and end-of-life recycling strategies (Angelaki et al., 2023; Goel et al., 2024).

4.6.2.2 Data Ethics and Responsible Innovation

This is the data on which this digitalization has reliance, and there are significant ethical responsibilities on usage of this data ethically. Companies are collected solicited and unsolicited data of their customers, analyze and leverage vast quantities of consumer information, they must be vigilant about how that collected data is handled. The privacy of data, the consent of customers for usage of this data and fairness must be embedded into all governance frameworks. Corporations need to ensure that the robust cybersecurity measures, transparent data polices, and full compliance with privacy regulations should be there.

4.6.2.3 Evolving Models of Corporate Governance

There is need to evolve the traditional corporate governance frameworks to keep pace with the speed, scale, and complexity of technological advancements. The board of directors must not only understand the financial and operational risks but also look into the emerging challenges of AI ethics, cybersecurity, environmental impact, and data governance. There is need to established effective global governance frameworks that include:

1. Establishing sustainability and digital ethics committees.
2. Integrating ESG performance into executive evaluation and incentives.
3. Instituting regular ethics and audits and risks assessments.
4. Promoting a company-wide culture of ethical innovation.

Here, corporate governance itself serves as the first line of accountability, ensuring that the strategic decisions, either in favor or against, reflect both business interests and broader societal responsibilities. As the digital transformation accelerates, governance structure must be dynamic, forward-looking, and have strong base in ethical leadership.

4.6.2.4 Stakeholder Engagement in the Digital Era

These are the digital platforms which have refined the relationship between the stakeholders and the corporations. Now consumers, employees, investors, regulators, and NGOs have more power and visibility than ever before. Corporations which have embraced these shifts are not only facing threats but also have an opportunity of partnerships for long-term success. The effective engagement of stakeholders involves in more than passive consultation and it requires active listening, transparency, and collaboration. The usage of real-time data analytics can help the corporations to tailor the sustainability efforts and initiatives to specific consumer needs and expectations, while digital interfaces enable feedback loops that foster the co-creation of polices and products. Stakeholders always have keen concerns that corporations must adhere strong ESG standards that attract the investors to give priority to those, which are leading the sustainability in capital allocation, consumers support brands and advocacy groups. In response business must embed stakeholder voices into the form of decision-making from the design of the product to ultimately at the end of supply chain policies.

4.7 DISCUSSION QUESTIONS WITH ANSWERS

1. *What are the core technologies driving Industry 4.0, and how do they impact traditional industries?*

Answer: Industry 4.0 is driven by AI, IoT, big data analytics, Cloud Computing, Cyber-Physical Systems (CPS), and advanced automation. These technologies enable real-time decision-making, predictive maintenance, and automation of repetitive tasks, transforming traditional industries into smart, adaptive, and efficient systems. Examples include smart manufacturing using robotics and AI-powered demand forecasting in supply chains.

2. *How does human capital play a central role in the success of Industry 4.0?*

Answer: Human capital is essential for the transition to Industry 4.0. Success depends on developing digital competencies, promoting lifelong learning, and addressing the skills gap. A digitally literate workforce enables innovation and adaptability. Social inclusion, equitable access to education, and reskilling programs are critical to ensure no one is left behind.

3. *Discuss the three pillars of sustainability in the context of Industry 4.0. How are they interconnected?*

Answer: These pillars are interconnected—green technologies (environmental) can reduce costs (economic) while improving public trust (social).

- Environmental Sustainability: Through energy-efficient systems, resource optimization, and circular economy practices.
- Social Sustainability: By enhancing job satisfaction, workplace safety, inclusivity, and equitable access to digital tools and education.
- Economic Sustainability: Achieved via increased productivity, innovation, and operational resilience.

4. *What are the ethical considerations corporations must address in adopting digital technologies?*

Answer: Corporations must address:

- AI Ethics: Preventing algorithmic bias and ensuring fairness.
- Data Privacy: Complying with regulations like GDPR and gaining user consent.
- Transparency: Informing stakeholders how their data is used.
- Inclusivity: Ensuring automation does not lead to mass displacement and inequality.
- Ethical frameworks should be embedded into the entire decision-making and innovation process.

5. *What strategies can governments and organizations adopt to address the workforce challenges posed by Industry 4.0?*

Answer: Governments can develop national skills strategies, fund vocational education, and ensure social safety nets. Organizations can foster cultures of continuous learning, create reskilling programs, and align with SDGs. Public-private partnerships should design industry-aligned curricula to keep pace with technological demands.

6. *How can Industry 4.0 technologies contribute to environmental monitoring and climate action?*

Answer: Technologies such as satellite imagery, AI-powered analytics, IoT sensors, and drones allow for precise tracking of environmental changes, pollution, and natural resource usage. Predictive analytics can forecast environmental risks like floods or droughts, enabling proactive measures for sustainability.

7. *Evaluate how sustainability can be a strategic advantage in the digital age.*

Answer: Sustainability, when embedded into corporate strategies, enhances brand reputation, attracts investors, meets regulatory requirements, and promotes innovation. Companies like Tesla leverage AI and green technologies to build environmentally responsible yet profitable business models. Sustainability can drive customer loyalty and long-term competitiveness.

8. *What is meant by the term "Sustainability 4.0"?*

Answer: Sustainability 4.0 refers to the integration of sustainability principles with Industry 4.0 technologies. It emphasizes balancing innovation with ethical, social, and environmental responsibility. It uses tools like AI and IoT not just for profit but also for achieving ecological resilience, social equity, and economic inclusion.

REFERENCES

Aditya, K., Chakraborty, S., Dahire, A., Kumari, A., & Milanova, M. (2024). Integrating Blockchain, IoT, and AI in supply chain management. In *Blockchain, IoT, and AI Technologies for Supply Chain Management: Apply Emerging Technologies to Address and Improve Supply Chain Management* (pp. 237–264). Apress.

Ahmed, S., & Smith, E. (2023). The future of work: Adapting to remote and hybrid models. *Abbottabad University Journal of Business and Management Sciences, 1*(01), 1–12.

Ajekwene, K. K., Aigbokhan, E., Akindele, O., Yibowei, M., Momoh, F., & Ugonna, U. (2022). Electronic waste (e-waste): Sources, proliferation, effects & management in developing nations. *IOSR Journal of Engineering, 12*, 12–27.

Aljohani, A. (2023). Predictive analytics and machine learning for real-time supply chain risk mitigation and agility. *Sustainability, 15*(20), 15088.

Angelaki, M. E., Bersimis, F., Karvounidis, T., & Douligeris, C. (2023). Exploring the awareness outcomes of educating ICT students on the environmental implications of e-waste recycling and energy consumption in data centers. *IEEE Transactions on Education, 67*(1), 143–152.

Ansari, Mohammad Sultan Ahmad, Ms. Xiaoyang Zhou, Phd. Available at SSRN: https://ssrn.com/abstract=4978687 or http://dx.doi.org/10.2139/ssrn.4978687

Ansari, I., Barati, M., Sadeghi Moghadam, M. R., & Ghobakhloo, M. (2023). An Industry 4.0 readiness model for new technology exploitation. *International Journal of Quality & Reliability Management, 40*(10), 2519–2538.

Arvind Mallik, D. M. (2024). Promoting green innovation in India's Industry 4.0: paving the way to a net-zero economy. In *Net Zero Economy, Corporate Social Responsibility and Sustainable Value Creation: Exploring Strategies, Drivers, and Challenges* (pp. 203–221). Springer Nature Switzerland.

Bakker, K. (2024). *Gaia's Web: How Digital Environmentalism Can Combat Climate Change, Restore Biodiversity, Cultivate Empathy, and Regenerate the Earth.* MIT Press.

Bandara, N. (2023). The role of green technology and AI in the development of the smart and sustainable town in ASIA: Singapore. *Proceedings of the 2nd international conference on sustainable & digital business.*

Blithe, S. J. (2023). Work-life balance and flexible organizational space: Employed mothers' use of work-friendly child spaces. *Management Communication Quarterly, 37*(4), 821–845.

Bykova, V., Khasanova, M., & Polonkoeva, F. (2024). The role of human capital in economic development: An analysis of factors contributing to economic growth. *Reliability: Theory & Applications, 19*(SI 6 (81)), 1676–1682.

Challoumis, C. (2024). The imperative of skill development in an AI revolution. *XIX international scientific conference.* London. Great Britain.

Cheng, Y., Masukujjaman, M., Sobhani, F. A., Hamayun, M., & Alam, S. S. (2023). Green logistics, green human capital, and circular economy: The mediating role of sustainable production. *Sustainability, 15*(2), 1045.

Chirita, M., & Sarpe, D. A. (2024). *Navigating the challenges and ethics of AI in shaping the future of work for sustainable Industry 4.0. Annals of "Dunarea de Jos" University of Galati: Fascicle I. Economics & Applied Informatics, 30*(3), 275–285. https://doi.org/10.35219/eai15840409453

Chukwunweike, J., Anang, A. N., Adeniran, A. A., & Dike, J. (2024). Enhancing manufacturing efficiency and quality through automation and deep learning: Addressing redundancy, defects, vibration analysis, and material strength optimization. *World Journal of Advanced Research and Reviews, 23*(3), 1272–1295.

Cirillo, V., Fanti, L., Mina, A., & Ricci, A. (2023). The adoption of digital technologies: Investment, skills, work organisation. *Structural Change and Economic Dynamics, 66*, 89–105.

Da Roit, B., & Iannuzzi, F. E. (2023). One of many roads to Industry 4.0? Technology, policy, organisational adaptation and worker experience in 'Third Italy' SMEs. *New Technology, Work and Employment, 38*(2), 252–271.

Dean, S. A., & East, J. I (2019). Soft skills needed for the 21st century workforce. *International Journal of Applied Management and Technology, 18*(1), 17–32.

de Oliveira, A. R., Pacheco, F. A. L., Fernandes, L. F. S., & da Costa, A. M. (2023). Trends of potentially illegal deforestation over protected areas of Rio das Velhas Basin, Brazil. *Land Degradation & Development, 34*(15), 4649–4666.

El Merroun, M., Bartók, I. J., & Alkhlaifat, O. (2022). Industry 4.0 technologies' effects on environmental sustainability-a systematic literature review. *Journal of Manufacturing Engineering, 17*(4), 132–152.

Elzawi, A., Kenan, T., Wade, S., & Pislaru, C. (2012). Bridging the digital divide and enhancing the quality of engineering research in Libyan universities. In *6th Conference on quality in Middle East, 2012, Hamdan Bin Mohammed University, Dubai.*

Emma, L. (2024). *The strategic use of big data analytics for scenario planning and risk management.* Retrieved December 3, 2024, from https://www.researchgate.net/publication/386375950_The_Strategic_Use_of_Big_Data_Analytics_for_Scenario_Planning_and_Risk_Management

Enyejo, J. O., Fajana, O. P., Jok, I. S., Ihejirika, C. J., Awotiwon, B. O., & Olola, T. M. (2024). *Digital twin technology, predictive analytics, and sustainable project management in global supply chains for risk mitigation, optimization, and carbon footprint reduction through green initiatives. International Journal of Innovative Science and Research Technology*, *9*(11), Article 1344. https://doi.org/10.38124/ijisrt/IJISRT24NOV1344

Goel, A., Masurkar, S., & Pathade, G. R. (2024). An overview of digital transformation and environmental sustainability: Threats, opportunities, and solutions. *Sustainability*, *16*(24), 11079.

Gupta, S. D. (2024). Building trustworthy AI: Proactive guardrails and reactive moderation for scalable governance. *International Journal of Innovative Research in Science Engineering and Technology*. https://doi.org/10.15680/ijirset.2024.1312181

Hayat, A., Shahare, V., Sharma, A. K., & Arora, N. (2023). Introduction to Industry 4.0. In S. Namasudra & K. Akkaya (Eds.), *Blockchain and Its Applications in Industry 4.0* (pp. 29–59). Springer.

Hayyat, A., Khan, H. A., Shafiq, M. A., & Ziaullah, M. (2023). Does Employee's green commitments and green attitude mediate the relationship of green HRM practices and pro-environmental behaviour: Evidence from TEVT Sector Southern Punjab, Pakistan. *Journal of Social Sciences Review*, *3*(2), 190–210.

He, X., Li, Y., Liu, S., Xu, T., Chen, F., Li, Z., Zhang, Z., Liu, R., Song, L., & Xu, Z. (2023). Improving regional climate simulations based on a hybrid data assimilation and machine learning method. *Hydrology and Earth System Sciences*, *27*(7), 1583–1606.

Hrytsevich, S., & Tahir, M. A. (2025). Application of the business ecosystem approach to the transformation of business models in industry: The example of the light industry transformation. In A. Hassan, S. Ullah, S. E. Barykin, & E. de la Poza (Eds.), *Multidisciplinary Analysis of Digital Transformation and Global Market Dynamics* (pp. 25–42). IGI Global.

Hu, Y., Jia, Q., Yao, Y., Lee, Y., Lee, M., Wang, C., Zhou, X., Xie, R., & Yu, F. R. (2024). Industrial Internet of Things intelligence empowering smart manufacturing: A literature review. *IEEE Internet of Things Journal*, *11*(11), 19143–19167.

Jeffers, P. I. (2010). Embracing sustainability: Information technology and the strategic leveraging of operations in third-party logistics. *International Journal of Operations & Production Management*, *30*(3), 260–287.

Joseph, L. P., Sherin, L., Ridha, Z. T., Mathew, S., Jose, S. K., Sebastian, T., Archana, L., Veena, R., Sreelatha, K., & Cyriac, J. (2025). Environment sustainability: Climate change mitigation, renewable resources, and biodiversity preservation aided by IoT and AI. In P. Aguilar-Zárate, M. L. Chávez-González, A. R. Abraham, & A. K. Haghi (Eds.), *The Food-Energy-Water Triangle for Sustainable Development* (pp. 269–291). Apple Academic Press.

Kess-Momoh, A. J., Tula, S. T., Bello, B. G., Omotoye, G. B., & Daraojimba, A. I. (2024). Strategic human resource management in the 21st century: A review of trends and innovations. *World Journal of Advanced Research and Reviews*, *21*(1), 746–757.

Khan, S. S., Te Lintelo, D., & Macgregor, H. (2023). Framing 'slums': Global policy discourses and urban inequalities. *Environment & Urbanization*, *35*(1), 74–90.

Knell, M. (2024). Technological revolutions and the co-evolution of green and digital transformation processes: Tapping the potential of diversified energy networks. *Journal of Innovation Economics & Management*, *3*(3), I–XXV.

Kopeinig, J., Woschank, M., & Olipp, N. (2024). Industry 4.0 technologies and their implications for environmental sustainability in the manufacturing industry. *Procedia Computer Science*, *232*, 2777–2789.

Lakshmi, S., Durgude, U., & Chitra, V. (2024). The role of upskilling and reskilling for talent transformation in the era of AI; theoretical framework and future research directions. In *Advanced Trends in Multidisciplinary Research and Innovation* (pp. 198). Weser Books.

Li, H., Khattak, S. I., Lu, X., & Khan, A. (2023). Greening the way forward: A qualitative assessment of green technology integration and prospects in a Chinese technical and vocational institute. *Sustainability, 15*(6), 5187.

Logeswaran, K., Savitha, S., Suresh, P., Prasanna Kumar, K., Gunasekar, M., Rajadevi, R., Dharani, M., & Jayasurya, A. (2024). Unifying technologies in Industry 4.0: Harnessing the synergy of Internet of Things, big data, augmented reality/virtual reality, and blockchain technologies. In M. R. AL-Refaey, A. K. Tyagi, A. S. AL-Malaise AL-Ghamdi, & S. Kukreja (Eds.), *Topics in Artificial Intelligence Applied to Industry 4.0* (pp. 127–147), Wiley.

Luz, A. (2025). *The role of digital platforms in enhancing the market reach of small businesses in rural America*. https://doi.org/10.20944/preprints202502.1162.v1

Magara, T., & Zhou, Y. (2024). Internet of Things (IoT) of smart homes: Privacy and security. *Journal of Electrical and Computer Engineering, 2024*(1), 7716956.

Majka, M. *The power of predictive analytics and machine learning models.*

Mayer, A., Greif, L., Häußermann, T. M., Otto, S., Kastner, K., El Bobbou, S., Chardonnet, J. R., Reichwald, J., Fleischer, J., & Ovtcharova, J. (2025). Digital twins, extended reality, and artificial intelligence in manufacturing reconfiguration: A systematic literature review. *Sustainability, 17*(5), 2318.

Mazhar, T., Irfan, H. M., Haq, I., Ullah, I., Ashraf, M., Shloul, T. A., Ghadi, Y. Y., Imran, & Elkamchouchi, D. H. (2023). Analysis of challenges and solutions of IoT in smart grids using AI and machine learning techniques: A review. *Electronics, 12*(1), 242.

Mishra, A. A., & Pathak, D. K. (2024). Industry 4.0 technologies adoption and sustainability integration in human resource management: An analysis using extended TOE framework and TISM. *IEEE Transactions on Engineering Management, 77*, 14688–14703.

Muchenje, C., Tapera, M. C., Katsvairo, H. T., & Mugoni, E. (2023). Green marketing strategies and consumer behavior: Insights for achieving sustainable marketing success. In R. Masengu, S. Bigirimana, O. T. Chiwaridzo, B. Ruzive, & C. Blossom (Eds.), *Sustainable Marketing, Branding, and Reputation Management: Strategies for a Greener Future* (pp. 465–484). IGI Global.

Napoli, F. (2023). Corporate digital responsibility: A board of directors may encourage the environmentally responsible use of digital technology and data: Empirical evidence from Italian publicly listed companies. *Sustainability, 15*(3), 2539.

Narkhede, G., Dohale, V., & Mahajan, Y. (2024). Darker side of Industry 4.0 and its impact on triple-bottom-line sustainability. *Sustainable Development, 32*(6), 5999–6016.

Osuya, N. U., & Cheta-Maclean, S. (2024). The emotional intelligence of leaders and strategic planning in public universities. *International Journal of Institutional Leadership, Policy and Management, 6*(3), 298–314.

Parashar, B., Sharma, R., Rana, G., & Balaji, R. (2023). Foundation concepts for Industry 4.0. In A. Nayyar, M. Naved, and & R. Rameshwar (Eds.), *New Horizons for Industry 4.0 in Modern Business* (pp. 51–68). Springer.

Pozzi, R., Rossi, T., & Secchi, R. (2023). Industry 4.0 technologies: Critical success factors for implementation and improvements in manufacturing companies. *Production Planning & Control, 34*(2), 139–158.

Qudus, L. (2025). Leveraging artificial intelligence to enhance process control and improve efficiency in manufacturing industries. *International Journal of Computer Applications Technology and Research, 14*(02), 18–38.

Ramasamy, R., Gopi, R. S., & Palanisamy, R. (2024). E-waste management in the digital era: A sustainable computing approach. *IEEE Potentials, 42*, 2–8.

Rane, N., Choudhary, S., & Rane, J. (2023). *Artificial intelligence (AI) and Internet of Things (IoT)-based sensors for monitoring and controlling in architecture, engineering, and construction: Applications, challenges, and opportunities*. SSRN Electronic Journal. https://doi.org/10.2139/ssrn.4642197

Rani, G. M., Pathania, D., Umapathi, R., Rustagi, S., Huh, Y. S., Gupta, V. K., Kaushik, A., & Chaudhary, V. (2023). Agro-waste to sustainable energy: A green strategy of converting agricultural waste to nano-enabled energy applications. *Science of the Total Environment, 875*, 162667.

Rasul, M. (2024). The digital revolution: How technology is shaping the global economy. In S. Shaheen, M. Bashir, & M. Khurrum Bhutta (Eds.), *Energy Crisis and Its Impact on Global Business* (pp. 256–287). IGI Global.

Ren, S., Cooke, F. L., Stahl, G. K., Fan, D., & Timming, A. R. (2023). Advancing the sustainability agenda through strategic human resource management: Insights and suggestions for future research. *Human Resource Management, 62*(3), 251–265.

Santos, L. B., Melo, F. J. C., Guimaraes Junior, D. S., Sobral, E. F. M., & Medeiros, D. D. (2023). Application of ISM to identify the contextual relationships between the sustainable solutions based on the principles and pillars of Industry 4.0: A Sustainability 4.0 model for law offices. *Sustainability, 15*(19), 14494. https://doi.org/10.3390/su151914494

Scheyvens, R. A., Movono, A., & Auckram, S. (2023). Pacific peoples and the pandemic: Exploring multiple well-beings of people in tourism-dependent communities. *Journal of Sustainable Tourism, 31*(1), 111–130.

Shabur, M. A. (2024). A comprehensive review on the impact of Industry 4.0 on the development of a sustainable environment. *Discover Sustainability, 5*(1), 97.

Shaheen, M. A., Hayyat, A., Aslam, S., & Shahbaz, M. A. (2023). Developing a culture of continuous learning for sustainable job performance: TEVT sector as learning organization, Pakistan. *Journal of Social Sciences Review 3*(2), 817–827.

Shaikh, M. K. Digital Twins and Engineering Education: Current Status.

Shan, Z., & Wang, Y. (2024). Strategic talent development in the knowledge economy: A comparative analysis of global practices. *Journal of the Knowledge Economy, 15*(4), 19570–19596.

Single, M., Bruhin, L. C., Schütz, N., Naef, A. C., Hegi, H., Reuse, P., Schindler, K. A., Krack, P., Wiest, R., & Chan, A. (2023). Development of an open-source and lightweight sensor recording software system for conducting biomedical research: Technical report. *JMIR Formative Research, 7*, e43092.

Sołtysik, M., Tyrańska, M., Piwowar-Sulej, K., & Agustina, T. S. (2024). *Sustainable Human Resource Management: Strategy, Organizational Innovation and Leadership in Industry 5.0.* Taylor & Francis.

Tursunbayeva, A., Pagliari, C., Di Lauro, S., & Antonelli, G. (2022). The ethics of people analytics: Risks, opportunities and recommendations. *Personnel Review, 51*(3), 900–921.

Ullah, S., Kukreti, M., Sami, A., & Shaukat, M. R. (2024). Leveraging technological readiness and green dynamic capability to enhance sustainability performance in manufacturing firms. *Journal of Manufacturing Technology Management, 36*(3), 714–730.

Usmani, U. A., Sulaiman, S., & Watada, J. (2024). Intelligent Integration of IoT and cyber-physical systems: Empowering the next generation of AI-enabled smart environments. *2024 8th international conference on computing, communication, control and automation (ICCUBEA).*

Vandenbogaerde, L., Verbeke, S., & Audenaert, A. (2023). Optimizing building energy consumption in office buildings: A review of building automation and control systems and factors influencing energy savings. *Journal of Building Engineering, 76*, 107233.

Van Doren, B. M., Lostanlen, V., Cramer, A., Salamon, J., Dokter, A., Kelling, S., Bello, J. P., & Farnsworth, A. (2023). Automated acoustic monitoring captures timing and intensity of bird migration. *Journal of Applied Ecology, 60*(3), 433–444.

Van Hoang, T. (2024). Impact of integrated artificial intelligence and Internet of Things technologies on smart city transformation. *Journal of Technical Education Science, 19*(Special Issue 01), 64–73.

Veile, J. W., Birkel, H. S., Schmidt, M. C., Rücker, M., Müller, J. M., Hartmann, E., & Voigt, K. I. (2021). Green and lean?–Understanding ecological and environmental implications in the light of Industry 4.0. *IOP conference series: Materials science and engineering. Volume 1196, International Conference on Resource Sustainability (ICRS 2021) 19th–23rd July 2021, Dublin, Ireland.*

Verma, P. (2024). Transforming supply chains through AI: Demand forecasting, inventory management, and dynamic optimization. *Integrated Journal of Science and Technology, 1*(3).

Wang, Z., Yu, L., & Zhou, L. (2024). Navigating the blockchain-driven transformation in Industry 4.0: Opportunities and challenges for economic and management innovations. *Journal of the Knowledge Economy, 16*(1), 1–43.

Zhao, G., Jones, P., Liu, S., Lopez, C., Dennehy, D., & Chen, X. (2023). Analysis of the drivers of Industry 4.0 technology deployment to achieve agri-food supply chain sustainability: A hybrid approach. *2023 IEEE international symposium on technology and society (ISTAS).*

5 The Role of Technology in Driving Organizational Sustainability

Muhammad Zulkifl Hasan, Muhammad Zunnurain Hussain, Sonia Umair, and Umair Waqas

5.1 INTRODUCTION

5.1.1 Overview of Technology's Role in Sustainability

Modern organizations experience critical transformation through the fundamental power of technology in their sustainability efforts. Organizations use modern technological solutions to reduce their environmental impact and optimize operational effectiveness because global environmental challenges increase in severity. The dual purposes of technology consist of offering direct environmental solutions while it functions as an enabler to transform business models and supply chains along with organizational culture toward sustainable practices. The real-time assessment and monitoring of environmental impact is made possible through technological improvements such as intelligent sensors alongside data-powered decision systems along with automation solutions (Alaimo & Maggino, 2020; Delgado-Ceballos et al., 2023). The role continues to expand quickly because organizations use digital capabilities to meet sustainability standards and customer expectations along with meeting regulatory needs (Waqas et al., 2025; Umair et al., 2025; Xiao and Su, 2022).

5.1.2 Importance of Sustainable Practices in Modern Organizations

Organizations treat sustainability as both essential for their business planning and fundamental to responsible corporate actions right now. Organizations now treat sustainability as their basic requirement for enduring performance and stability. The linked economy directly affects how well organizations succeed by managing all relationships between the natural environment social factors and financial performance. Organizations that put sustainability standards into their business model can handle threats better and save money while getting more customers and keeping investor support. Companies must follow sustainable practices because national and global rules mandate lower carbon emissions and better resource usage plus equality among all people. Technology makes sustainability measurement more precise and helps companies manage their resources better while developing new environmentally friendly solutions. The current market environment demands technology-supported

DOI: 10.1201/9781003611905-7

sustainability practices to become a strategic advantage for businesses (Carter et al., 2019; Zhang et al., 2020).

5.2 UNDERSTANDING ORGANIZATIONAL SUSTAINABILITY

5.2.1 Definitions and Scope of Sustainability in Business

For a business to work sustainably, it protects the environment while treating everyone justly and ensures lasting income for the future. Organizational sustainability works as a total approach that links environmental defense work with ethical standards to run business correctly under honest stakeholder contact and open reporting. A business promotes sustainability when it provides immediate benefits to customers and prepares for future generations to meet their needs. Our organization wants to handle resources more effectively alongside lowering carbon emissions and energy needs to better serve both our staff members and external communities. Each business operation affects sustainability at multiple points from selecting vendors to manufacturing and selling the product. The organization must define sustainability as its primary guide to manage business risks and make proper management changes to meet worldwide standards and customer requirements (Alaimo & Maggino, 2020; Delgado-Ceballos et al., 2023).

5.2.2 Key Metrics and Goals in Sustainable Development

Companies track their sustainability progress using selected performance indicators that show their results from ESG views of planet and people. Organizations regularly track their environmental performance using main indicators such as carbon emissions, energy use, water usage, waste output, and emission cuts. Organizations monitor worker health and happiness together with the success of their diversity programs and employee participation in local communities. They also assess their employment standards and conduct. Governance metrics show how transparent a company is while measuring whether it follows laws properly using ethical suppliers along with its range of board members (Alaimo & Maggino, 2020; Narimissa et al., 2020). These performance indicators match with standards from organizations like GRI, SASB, and the SDGs of the UN. Organizations use clear sustainability objectives with time limits to show their progress while gaining responsible investor interest and revealing their sustainable work. Through technology organizations can measure performance results accurately and take timely actions to improve their performance.

5.3 TECHNOLOGICAL INNOVATIONS SUPPORTING SUSTAINABILITY

5.3.1 Digital Transformation and Green Technologies

Green technologies are transforming how organizations and how the world interact with one another and this transformation is known as digital transformation. Green technologies include innovations aimed at lessening environmental damage, such as renewable energy systems, energy-efficient equipment, sustainable materials, etc. These green technologies are digitally transformed by integrating them into business operations using tools such as automation, artificial intelligence (AI), and the cloud computing. As a result of this integration, we create better systems that lessen waste and emission and manage the use of the resources more efficiently. For example, digital platforms can also automate energy management in buildings to use energy optimally without a human being. Digital transformation and green innovation synergy makes organizations more agile and equally more efficient, and environmentally responsible and thus, forms the basis of a sustainable digital economy (Guo et al., 2021; Hannan et al., 2021; Xiao and Su, 2022.

5.3.2 Impact of AI and Machine Learning in Sustainability

Machine learning concepts and AI are changing the way organizations approach sustainability. These technologies analyze huge amounts of data and recognize patterns, predict results, and work more efficiently with minimum involvement of the human factor. In sustainability contexts, AI is utilized for forecasting energy consumption, equipment failures, and environmental scenarios of different organizational decisions. ML models can suggest potential areas of improvement in an organization's supply chain, best approaches toward sustainability, and even create sustainability reports. They are also used for climate modeling and stock, difference environment status and pollution detection to allow organizations and governments to act immediately on such threats (Hannan et al., 2021). With this kind of development, the accuracy of these systems is also being able to give better understanding to the decision-makers in a bid to enable them to act appropriately and for the longest time. But the energy intensity of the AI infrastructure applied must be supervised to avoid the situation when the solutions being implemented become a source of the mentioned problems (Dang and Serajuddin, 2020).

5.3.3 Role of IoT in Monitoring and Reducing Environmental Impact

The environmental monitoring and management in organizations rely tremendously on the Internet of Things (IoT). IoT-embedded sensors in the equipment, infrastructure, and products allow the gathering of continuous data over various variables like temperature, humidity, emissions, energy usage, and waste levels. This in real-time data allows organizations to see how wasting, responses, and systems are working to give you minimal impact on the environment. According to GSMC, an example of smart HVAC systems would be ones that adjust the heating and cooling based on occupancy and weather, and smart water meters that can detect leaks almost immediately to eliminate wastage. Predictive maintenance that conserves resources is supported by IoT and reduces downtime and prolongs equipment life. When coupled with AI and data analytics – IoT becomes a potent force in the creation of intelligent, responsive, and environmentally aware organizational ecosystems (Alotaibi, 2019).

5.3.4 Cloud Computing and Energy Efficiency

The computing resources allows organizations to use their computing power flexibly and dynamically which in turn reduces their IT infrastructure's carbon emissions (Guo et al., 2021). Organizations achieve environmental benefits when they move their services into cloud infrastructure because they minimize their server requirements and reduce repeated systems and improve power utilization. Major cloud service providers use their investments toward renewable energy-powered data centers that facilitate their customers to cut their carbon footprints. Enterprise cloud solutions provide users with measurement tools to track their energy consumption together with simulation capabilities along with complete sustainability data management features throughout their global network of operations. Virtualization of services while automating resource scaling allows businesses to consume energy only during needed periods thus avoiding traditional IT systems' excessive overprovisioning. Digital infrastructure growth allows the cloud to become a sustainable solution which maintains both environmental responsibility and computing performance and adaptability.

5.4 SUSTAINABLE SUPPLY CHAIN MANAGEMENT

5.4.1 Leveraging Blockchain for Transparency and Accountability

Also, within the field of sustainable SCM, there is a tremendous opportunity in the use of blockchain technologies to enhance the transparency, accountability, or supply chain tracks in all tiers. The disruption of supply chains remains challenging, especially regarding the provenance of products or the occurrence of fraud, as modern supply chains are traditionally complex and obscure. Blockchain is of enormous benefit due to its decentralized and tamper-proof characteristic, which are ideal in the creation of a genuine delivery trail. It would also be possible to use blockchain in tracking the origin of raw materials in an organization, ensure the company adheres to environmental standards, as well as ensure that sustainability certifications are genuine. For instance, a food-producing company can track a food commodity till the farm from where it was produced and can ensure that the food was produced through sustainable farming methods. Smart contracts also feature in the system to enhance compliance and reporting within the environment, and they execute actions once a defined environmental or ethical compliance or violation occurs. Apart from establishing trust between consumers and assurance of stakeholders, blockchain enables transparent supply chain to reduce cost and liabilities on operations (Alotaibi, 2019; Carter et al., 2019).

5.4.2 Optimizing Logistics and Reducing Carbon Footprint

Logistics is an essential principle of supply chain management, and an important aspect of sustainable supply chain management is the efficient use of a technology. Transportation and warehousing contribute a huge chunk toward the supply chain emissions hence, there is potential for making big changes to these systems and its functions will create a great impact to sustainability. Automated logistics systems promptly incorporate real-time data, scripted algorithms, or even machine learning to employ more efficient and fuel-saving routes as well as to limit empty running. GPS tracking, telematics, and dynamic scheduling enhance the use of this resource and make it possible to allocate resources to the right place. In warehouses, automation systems and robots help in proper handling and arranging of stocks, reduced on power consumption as well as human interferences. Electric and hybrid vehicles are gradually shifting and replacing the conventional diesel vehicles, including trucks, which will help in reduction of emission levels. Also, the logistics networks' digital copies help to predict various transportation situations and choose the environmentally friendly ones and estimate their emissions of carbon. Thus, it is possible to achieve a balance between emissions reduction, organization performance, reliability, and speed in customer service (Carter et al., 2019; Narimissa et al., 2020; Yadav et al., 2020).

5.4.3 Case Studies in Sustainable Supply Chains

There are numerous examples of the concepts of building sustainable supply chains through case of real-life implementation put forward by several pioneers in the field. Unilever is a notable example, as it has implemented a significant effort on the use of digital technologies to get a clear and adequate picture of its supply chain and the sustainability factors that are present in their global operations. The company follows the path of using technology, such as blockchain for supply chain tracking of raw materials and the AI for suppliers' conformity and identification of the shortest routing for delivery and its environmental impact. Sustainable clothing companies utilizes IoT and data analytics to calculate the sustainability metrics of its produce from the acquisition of nature-derived material such as the organic cotton right from the farmers all along the production channels, to utilization, and end-of-life disassembly and recycling. The supply chain of the brand is well traced to ensure that it complies with the set environmental as well as the social standards. In the same way, Maersk, one of the largest container shipping companies, has incorporated the use of blockchain and AI to make trade documents electronic and to enhance the flow of containers and reduce on fuel consumption and time consumed. These case studies clearly demonstrate that emphasis on sustainable ways to use technologies does not only help companies to meet organizational goals, but it also serves to build and solidify the company's reputation and allows systems and processes to pass regulatory compliance checks and balances, thereby reaping competitive advantages (Dzikriansyah, 2023; Khan, 2022; Qalati, 2022; Sheng et al., 2023).

5.5 THE ROLE OF DATA AND ANALYTICS IN SUSTAINABILITY

5.5.1 Data-Driven Decision-Making for Sustainability

Information has been considered as the cornerstone of sustainable development measures. Every activity nowadays can be tied to a figure and, therefore, selecting, achieving, and improving environmental goals must take this into account. Through such data gathering and analysis of processes throughout operations energy, emissions, waste, water, and others are established to provide the best strategic direction in mass intervening points. Managers do not use estimating or reports from the past; using real-time items such as dashboard and predictive analytics to monitor and tackle environmental issues. For instance, energy management platforms include information from IoT systems that operate with manufacturing plants and offices to control the use of energy. Likewise, information obtained from suppliers may help make sustainable procurement decisions to avoid high impact and unethical suppliers. Since sustainability has been taking a central position in the management of organizations, the application of data analysis in managerial decision-making guarantees that environmental objectives are evidence-based (Lucivero, 2020; Tamym et al., 2023).

5.5.2 Predictive Analytics for Resource Optimization

Predicting is a branch of analytical activity, belonging to data science, which helps to optimize resource management by referring to the data of past and present periods. In sustainability areas, forecasting models are employed to estimate the energy and water requirements, equipment breakdowns, and waste generation, that is, to prevent particular occurrences, and minimize resource usage and negative impacts on the environment. For instance, in agriculture, predictive analytics can predict the probable climatic conditions and type of soil on farms and in turn inform farmers on when to apply water and fertilizers. In industrial sectors these models can estimate on when certain machinery might be about to fail and servicing is done to avoid energy wastage and expensive breakdowns. Supply chain also gets the benefit of predictive insights as an organization can predetermine when there might be disruptions, increase in demand, or shortage of materials for procurement processes. This not only improves the efficiency of the process but also minimizes risk by giving organizations better tools with which to forecast more effectively, as well as allocate better in an effort to meet organizational demands (Tamym et al., 2023).

Figure 5.1 shows the interdependencies between sustainability metrics where Cost Savings share 0.77 positive correlation with Carbon Footprint Reduction and Employee Engagement while having a relationship of 0.73 with Operational Efficiency Improvement. The square cells show the Pearson correlation coefficient values which span from –1 to 1 between metrics pairs. A good positive value in the correlation analysis indicates that the relationship between variables is strong. The relationship between Operational Efficiency Improvement and Cost Savings and Employee Engagement reaches the highest level at 0.77 and 0.73, respectively, although efficiency gains drive both financial impacts and employee satisfaction. The acceptance of sustainable practices demonstrates significant impacts on Cost Savings (0.75) besides showing strong correlations with Employee Engagement (0.75). The associations of Carbon Footprint Reduction with assessment metrics are strong, yet the relationship between this measure and Operational Efficiency (0.59) stands out as the weakest among all tested connections. The chart demonstrates how sustainability programs establish connections between financial and environmental management with human capital improvements.

Figure 5.2 presents key sustainability data relating to Cost Savings and Carbon Footprint Reduction and Operational Efficiency throughout the start and end years of 2019 and 2023. All sustainability metrics advanced from 2019 to 2021 before Operational Efficiency reached its peak at 40% during the same period and Cost

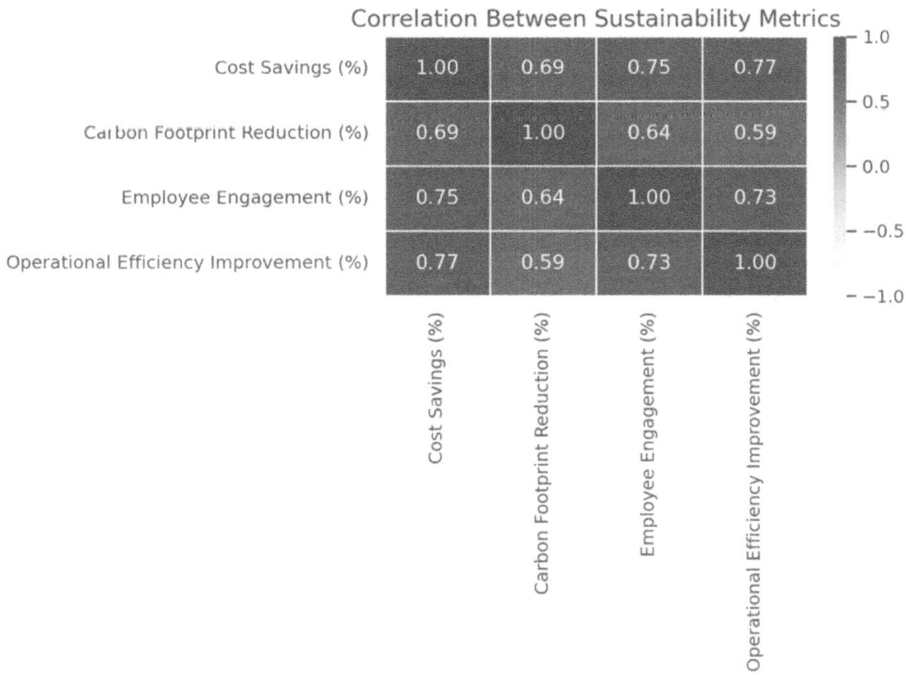

FIGURE 5.1 Correlation between sustainability. (Lastman, 2025)

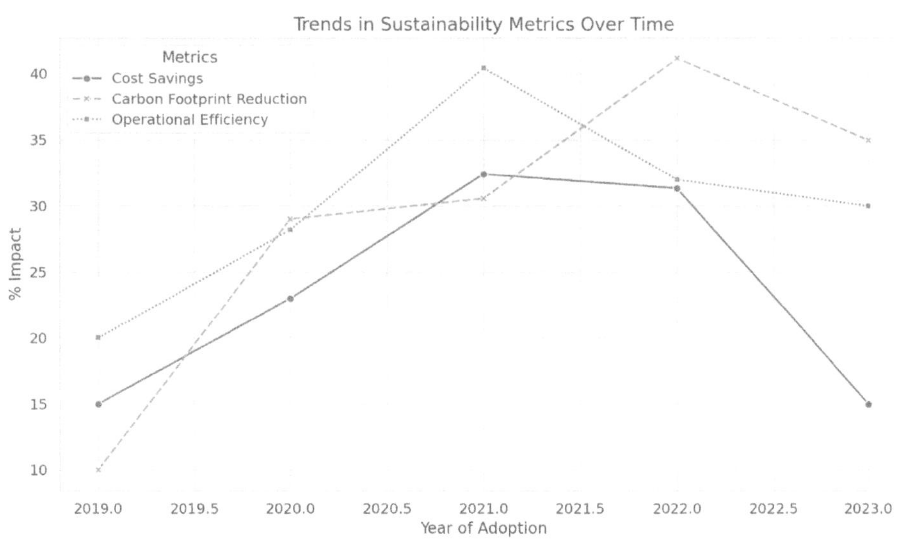

FIGURE 5.2 Trends in sustainability metrics. (Lastman, 2025)

Savings settled at 33%. The Carbon Footprint Reduction sector achieved the greatest advancement since its starting point of 10% in 2019 it activated to its top level of 41% in 2022. Post-2021 Cost Savings along with Operational Efficiency suffered declining slopes that produced a steep drop for Cost Savings reaching just 15% by 2023. The metric Carbon Footprint Reduction maintains its standing as the top performer since 2023 despite a momentary reduction in performance. Sustainability adoption proved beneficial for organizations initially yet building sustainable improvements in financial results together with operational achievements has proven complex for recent time periods.

Figure 5.3 shows the typical percentage cost reductions that result from using sustainability-related technologies. Blockchain technologies for energy together with smart grids systems lead the way regarding cost savings at 50% thus proving their substantial financial role in energy system optimization. The combination of smart agriculture and data center cooling solutions enables businesses to retain 45% and 40% savings thus demonstrating their strong abilities for operational enhancement and resource management. The cost-saving effects of AI in logistics together with cloud computing and electric vehicles amount to between 35% and 38% savings potential. AI waste reduction solutions and renewable energy systems and sustainable fabric development offer limited savings potential while reaching only 15% or less compared to organizational costs because the payback period extends beyond 15% decrease. The economic benefits among sustainability technologies demonstrate wide-ranging values according to this chart so businesses

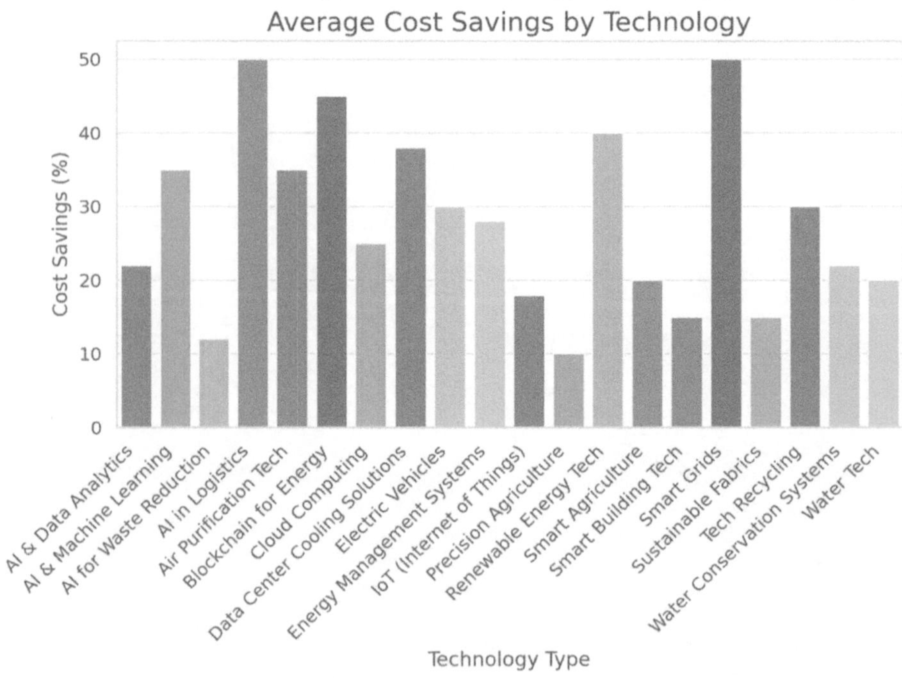

FIGURE 5.3 Average cost saving. (Lastman, 2025)

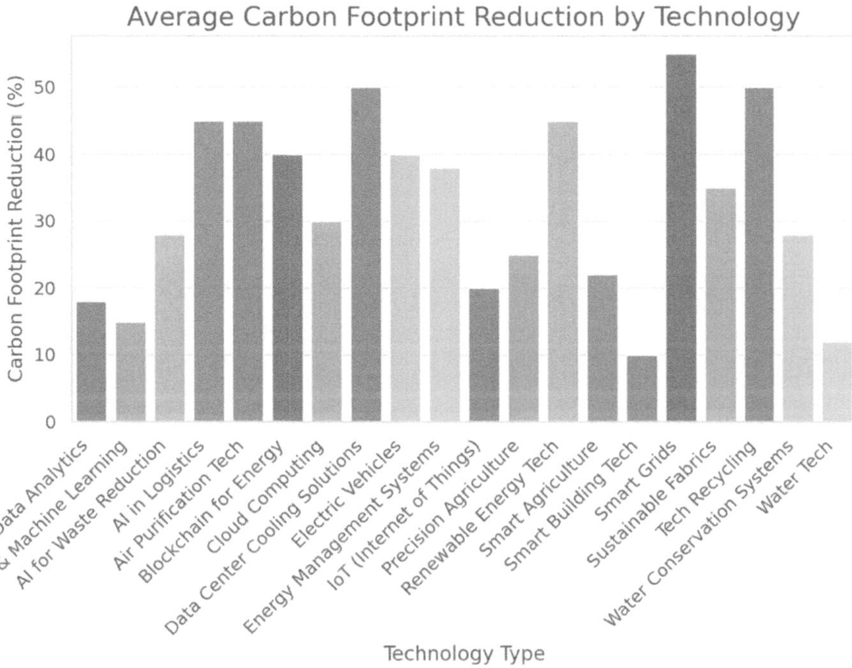

FIGURE 5.4 Average carbon reduction. (Lastman, 2025)

need to make purposeful technology choices that align with organizational targets and projected monetary returns.

Figure 5.4 shows the different sustainability technologies provide average carbon footprint reductions which are displayed in the attached bar chart. Sustainable fabrics provide the greatest reduction at 55% which is matched by data center cooling solutions along with water conservation technology at 50%. The environmental impact of air purification tech and blockchain for energy and smart agriculture results in reductions between 40% and 45%. Electric vehicles, energy management systems, and IoT show moderate performance, with reductions between 38% and 40%. The carbon emission reductions related to AI waste reduction together with smart tech and water solutions fall under 20% which represents their lower performance levels. The chart demonstrates substantial variations in how different technologies affect the environment thus organizations should evaluate environmental results together with operational outcomes when selecting sustainability investments.

Figure 5.5 shows that several sustainability technologies achieve average operational efficiency improvements reaching between 42% and 50%. AI logistics systems together with air purification technology demonstrate the greatest efficiency boost at 50% which demonstrates their significant capability to optimize operations and resource consumption. Various efficiency improvements for cloud computing and smart grids coupled with data center cooling solutions prove notable through values between 42% and 45%. The performance levels of electric vehicles and energy management systems and blockchain for energy technology fall between 38% and 44%.

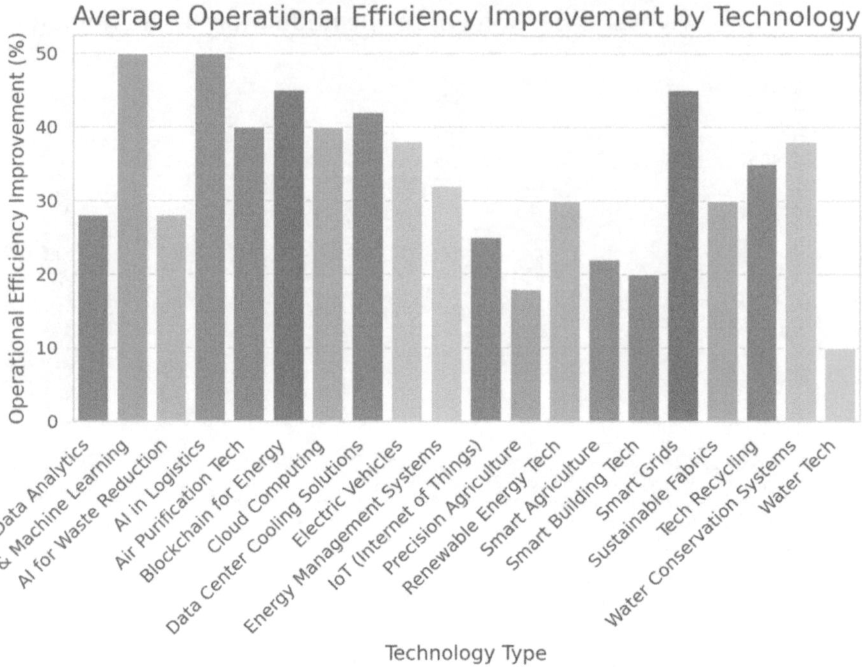

FIGURE 5.5 Operational efficiency by technology. (Lastman, 2025)

Renewable energy technology along with smart technology and water technology present efficiency gains that are below 20% in specific instances. The obtained data reveals that sustainability technologies affect operations differently because some innovations can produce substantial productivity gains, yet others need strategic incorporation to achieve their maximum effectiveness.

5.5.3 Environmental Impact Assessment through Big Data

The development of big data technology has brought the traditional Environmental Impact Assessment methodology into the modern era through better methods of data processing and manual analysis management. Organizations can achieve real-time control over ecological systems through environmental data acquisition conducted by bringing together satellites drones IoT devices and public databases. Advanced computational systems process collected data to run environmental analyses relating to industrial developments as well as infrastructure works together with new policies that affect air quality biodiversity and land usage and water management systems. Organizations benefit from big data to perform examinations through Environmental Impact Assessments while following environmental regulations to create proper mitigation strategies at fast speeds and high accuracy. The evaluation of ecosystem impact during project planning steps by construction businesses and emission analysis by energy producers through various fuel alternatives for decision-making serve as critical requirements for new construction developments. Implementing big data technology in environmental impact assessments grants organizations two primary benefits which include improved environmental safeguards and better community engagement that originates from scientifically verified impact evaluations (Lucivero, 2020).

5.6 CHALLENGES AND BARRIERS TO TECHNOLOGICAL ADOPTION FOR SUSTAINABILITY

5.6.1 Financial Constraints and Cost of Implementation

Small- and middle-sized companies plus organizations in developing areas struggle to use sustainability technology due to financial obstacles. Organizations need to pay for essential system components like sensors, AI equipment, and IT network changes at the start of rollout. The beginning expenses keep many organizations from starting even when they see future money savings. The benefits from sustainable technology investments appear in the future which makes them challenge to justify through traditional finance systems that Favor quick financial returns. Organizations need effective financial support and low-interest loans to accept sustainable technologies. The public and banking sectors should offer support including funding programs as well as tax benefits to partner with private entities and share sustainable innovation risks and rewards (Sahoo et al., 2023).

5.6.2 Regulatory and Compliance Issues

Running sustainability technology projects becomes harder due to different environmental regulations in various regions. Each region sets different rules which create regulation doubts and high administrative costs, especially for businesses that operate worldwide. Old laws that have not changed face difficulties in the modern world because new technologies do not have established legal foundations. Data privacy laws create problems if organizations need to collect IoT information for their environmental monitoring programs. Excessive reporting tasks take up valuable resources, especially for organizations that need display ability tools or expertise in sustainability data monitoring. The standardization of regulations and training programs that explain compliance rules help businesses deal with present challenges. Including compliance functions within digital systems helps automate reporting and keeps sustainability projects within legal requirements (Delgado-Ceballos et al., 2023).

5.6.3 Technological Limitations and Security Concerns

For sustainability solutions to progress, we need better technical solutions between systems and more compatible platform development. When older systems and modern technology do not connect well it leads to problems with combining data and operating information. Standardization of data parameters becomes impossible when hardware and software products from different suppliers must work together. In addition, concerns about cybersecurity pose significant barriers. Devices that talk to each other increase cybercriminal opportunities to attack digital systems. Poor cybersecurity in smart grid controls and network systems places your operations and environmental protection work at major risk. Security measures must be added during technology development to achieve good results because effective authority plus surveillance technology plus trained staff work best to safeguard operations. Digital systems function as secure and scalable formations with help from technology service partners and system engineers (Alotaibi, 2019).

5.7 CASE STUDIES AND REAL-WORLD APPLICATIONS

5.7.1 Case Study 1: Sustainable Manufacturing

Advanced manufacturing systems have a large number of benefits from the technology. Siemens leveraged digital twins AI and IoT technology in its plants to make use of power smarter and waste less while having better performance results as a leader. In case of low demand, data readings and predictions lead to production lines in order for energy to be saved instantly. The machine learning systems guard the materials (protect) by finding defects that can become a problem before they would become a problem and they discard less. The parts of used products can be refreshed and our modern technologies allow us to put these back into manufacture. Manufacturing better goods are even achieved by operations that rotate automatically at optimal energy levels to reach sustainability targets and save money. This case has showed that there is strong integration of smart technology that is effective in bringing profit and sustainability together (Sahoo et al., 2023; Sheng et al., 2023).

5.7.2 Case Study 2: Renewable Energy Utilization in IT

This is a prime example an IT organization can reconcile renewable energy and digital intelligence, in the form of data centers. By using AI systems developed by DeepMind to optimize power usage and cooling in real time, the company has promised to be entirely on carbon free energy from day to night. The systems forecast the availabilities of renewable energy and load workloads dynamically according to the availabilities of solar and wind power, so that the usage of solar and wind power is maximized. In addition to investing in grid decarbonization through purchasing renewable energy credits and funding clean energy projects, Google also invests in recent energy projects. These are not only the means by which the company cuts its impact on the environment, but they are the standards by which the entire tech industry should follow. In the case, data, AI and renewable energy have been harmonized to form a model for sustainable Digital Infrastructure (Guo et al., 2021; Tamilselvan & Sudhakar, 2024).

5.8 CONCLUSION

5.8.1 Summary of Key Takeaways

Technology plays a major role in building organizational sustainability while providing systems that improve resource efficiency and decrease environmental load plus give clear views of operations. Different modern technology solutions including AI, IoT, blockchain, and cloud computing help businesses understand their sustainability challenges better. Technological tools support better data-based choices about equipment maintenance and circular resource reuse which creates sustainable results for business and nature (Carter et al., 2019; Delgado-Ceballos et al., 2023). The pursuit of digital transformation faces obstacles like budget limits legal hurdles and technical vulnerabilities which call for well-planned teamwork among sectors and authorities as well as effective funding. Actual experiences in manufacturing and IT prove that by using technology correctly people transform sustainability needs from strict rules into new competitive strengths.

5.8.2 Future Directions for Technology-Driven Sustainability

Achieving environmental sustainability arises from integrating technology platforms with protection practices for the environment. The future of environmental sustainability will depend on quantum computing interfaces which use bioinformatics to operate decentralized energy infrastructure for boosting environmental management abilities (Hannan et al., 2021; Tamilselvan & Sudhakar, 2024). The blockchain, alongside cloud platforms, will merge into one sustainable operation system by organizations. Advanced standards of organizational norms together with consumer trust now enforce new laws to regulate the updated inventions. Success in this market demands businesses to use advanced technological equipment alongside trained personnel for preserving values that go beyond immediate monetary gains. The digital system development process will establish environment-focused technology which serves global human interests.

REFERENCES

Alaimo, L.S. and Maggino, F. (2020) "Sustainable development goals indicators at territorial level: Conceptual and methodological issues—the Italian perspective," Social Indicators Research, 147(2), pp. 383–419. https://doi.org/10.1007/s11205-019-02162-4.

Alotaibi, B. (2019) "Utilizing blockchain to overcome cyber security concerns in the internet of things: A review," IEEE Sensors Journal, 19(23), pp. 10953–10971. https://doi.org/10.1109/jsen.2019.2935035.

Carter, C.R. et al. (2019) "Sustainable supply chain management: Continuing evolution and future directions," International Journal of Physical Distribution & Logistics Management, 50(1), pp. 122–146. https://doi.org/10.1108/ijpdlm-02-2019-0056.

Dang, H.A.H. and Serajuddin, U. (2020) "Tracking the sustainable development goals: Emerging measurement challenges and further reflections," World Development, 127(104570), p. 104570. https://doi.org/10.1016/j.worlddev.2019.05.024.

Delgado-Ceballos, J. et al. (2023) "Connecting the sustainable development goals to firm-level sustainability and ESG factors: The need for double materiality," BRQ Business Research Quarterly, 26(1), pp. 2–10. https://doi.org/10.1177/23409444221140919.

Guo, Z. et al (2021) "Impact of energy storage on renewable energy utilization: A geometric description," IEEE Transactions on Sustainable Energy, 12(2), pp. 874–885. https://doi.org/10.1109/tste.2020.3023498.

Hannan, M.A. et al (2021) "Impact of renewable energy utilization and artificial intelligence in achieving sustainable development goals," Energy Reports, 7, pp. 5359–5373. https://doi.org/10.1016/j.egyr.2021.08.172.

Lastman (2025) Kaggle.com. https://www.kaggle.com/datasets/lastman0800/technology-sustainability.

Lucivero, F. (2020) “Big data, big waste? A reflection on the environmental sustainability of big data initiatives,” Science and Engineering Ethics, 26(2), pp. 1009–1030. https://doi.org/10.1007/s11948-019-00171-7.

Muhammad Akbar Dzikriansyah, I.M. (2023) “The role of green supply chain management practices on environmental performance: A case of Indonesian small and medium enterprises,” Cleaner Logistics and Supply Chain, 6, p. 12.

Khan, M. T., Idrees, M. D., Rauf, M., Sami, A., Ansari, A., & Jamil, A. (2022). Green supply chain management practices’ impact on operational performance with the mediation of technological innovation. Sustainability. https://doi.org/10.3390/su14063362

Narimissa, O., Kangarani-Farahani, A. and Molla-Alizadeh-Zavardehi, S. (2020) “Evaluation of sustainable supply chain management performance: Indicators,” Sustainable Development, 28(1), pp. 118–131. https://doi.org/10.1002/sd.1976.

Nureen, N., Liu, Exploring the technical and behavioral dimensions of green supply chain management: A roadmap toward environmental sustainability. Environmental Science and Pollution Research International, 29(42), 63444–63457. https://doi.org/10.1007/s11356-022-20352-5

Sahoo, S., Kumar, A. and Upadhyay, A. (2023) “How do green knowledge management and green technology innovation impact corporate environmental performance? Understanding the role of green knowledge acquisition,” Business Strategy and the Environment, 32(1), pp. 551–569. https://doi.org/10.1002/bse.3160.

Sheng, X., Chen, L., Yuan, X., Tang, Y., Yuan, Q., Chen, R., Wang, Q., Ma, Q., Zuo, J. & Liu, H. (2023) “Green supply chain management for a more sustainable manufacturing industry in China: A critical review,” Environment, Development and Sustainability, 25(2), pp. 1151–1183.

Qalati, S. A., Kumari, S., Soomro, I. A., Ali, R. and Hong, Y. (2022) “Green Supply Chain Management and Corporate Performance Among Manufacturing Firms in Pakistan,” Frontiers in Environmental Science, 10, Article 873837.https://doi.org/10.3389/fenvs.2022.873837

Tamilselvan, D. and Sudhakar, T.D. (2024) “Optimizing renewable energy utilization with high gain converters,” Renewable and Sustainable Energy Reviews, 191(114105), p. 114105. https://doi.org/10.1016/j.rser.2023.114105.

Tamym, L. et al (2023) “A big data analytics-based methodology for social sustainability impacts evaluation: A case study,” Procedia Computer Science, 220, pp. 32–39. https://doi.org/10.1016/j.procs.2023.03.007.

Umair, S., Waqas, U., Mrugalska, B., & Durrah, O. (2025) “Green human resource management, green environmental passion, pro-environmental behaviour, and sustainable environmental performance. A sequential mediation analysis,” Economics and Environment, 95(4), pp. 927–927. https://doi.org/10.34659/eis.2025.95.4.927

Waqas, U., Umair, S., Mrugalska, B. & Ali Sulaiman, M. A. B. (2025) “Twin transformation: The role of digitalization with artificial intelligence, big data and circular supply chain in circular economy performance for sustainability,” Equilibrium. Quarterly Journal of Economics and Economic Policy, 20(2), pp. 613–645. https://doi.org/10.24136/eq.3350

Xiao, D. and Su, J. (2022) “Role of technological innovation in achieving social and environmental sustainability: Mediating roles of organizational innovation and digital entrepreneurship,” Frontiers in Public Health, 10, p. 850172. https://doi.org/10.3389/fpubh.2022.850172.

Yadav, G. et al (2020) “A framework to achieve sustainability in manufacturing organisations of developing economies using industry 4.0 technologies’ enablers,” Computers in Industry, 122(103280), p. 103280. https://doi.org/10.1016/j.compind.2020.103280.

Zhang, Y. et al (2020) “Critical success factors of green innovation: Technology, organization and environment readiness,” Journal of Cleaner Production, 264(121701), p. 121701. https://doi.org/10.1016/j.jclepro.2020.121701.

Part III

Supply Chain and Logistics Transformation

6 Establishing Resilient and Sustainable Supply Chains in the Digital Transformation Era

Devarajanayaka Muniyanayaka, V.P. Sriram, Zeeshan Asim, and Thirupathi Manickam

6.1 INTRODUCTION TO RESILIENT AND SUSTAINABLE SUPPLY CHAINS

In the ever-changing world of international business, resilience and sustainability are now recognized as fundamental building blocks in the way we design and manage our contemporary supply chains. It's not an exaggeration to here use the word "disruption" because the coming together of so many global shocks—from climate change through pandemics and from geopolitics through digital transformation—has served to demonstrate the fragility of the linear supply chain and the neat network diagrams we all like to draw. As a result, companies are moving increasingly toward robust sustainable Agricultural Supply Chains (AGRSC) to resist exogenous shocks for long periods while promoting environmental and social responsibility. With regard to supply chains, resilience is the capability to anticipate, prepare for, respond to, and recover from a disruptive event. It means creating resiliency throughout procurement, production, distribution, and logistics. Sustainability, however, broadens this concept including environmental protection, ethical sourcing, Circular Economy (CE)-based principles, and stakeholder engagement. In its totality, these two dimensions constitute the core of creating continuity, competition, and corporate responsibility in an unstable global market. They are critical as they mitigate operational risks, improve client satisfaction, and guarantee compliance with regulations. Ivanov and Dolgui (2022) indicate also that the resilient supply chain can sustain operations during a crisis without loss of function, attenuation of performance, and ethical compromise. Concomitantly, sustainability-based strategies, including carbon foot- print minimization, zero-waste generation, and renewable energy utilization, are becoming critical with respect to brand image and investor expectations (Bag et al, 2023).

In today's digital world, several digital technologies, including Artificial Intelligence (AI), blockchain, Internet of Things (IoT), cloud, and advanced analytics, are changing the game in supply chains. These capabilities enable organizations to

DOI: 10.1201/9781003611905-9

forecast disruptions, inventory segmentation, product visibility throughout the value chain, and in the minute decision-making with real time. For example, blockchain provides end-to-end transparency and traceability in supply chains; this enhances both resilience and sustainability through the prevention of fraudulent activities, reduction of waste, and compliance with environmental laws (Kumar et al., 2024). Second, supply chain management (SCM) paradigms are shifting from being cost-centric to value-based. Organizations have moved beyond operational efficiency, and have increasingly been talking about agility, transparency, and collaboration. This shift is also stimulated by consumers seeking ethical and green products in the marketplace, leading companies to formulize their businesses in terms of social and environmental accountability. The amalgamation of CE principles such as reuse, remanufactures, and recyclable has evolved as an imperative approach while structuring sustainable supply chain frameworks (Zhang et al., 2025). Although no one can predict the future, what we have learned from this pandemic is that robust and sustainable supply chains are no longer a nice-to-have, but a strategic necessity. In an environment of ever more complex risks in an interconnected global economy, organizations need to adopt holistic supply chain strategies that are both flexible and sustainable. It is digital technologies that enable the dawning of this era, providing the capability to create networks that are agile, transparent, and that care about the future and society.

6.2 DRIVERS OF DIGITAL TRANSFORMATION IN SUPPLY CHAIN MANAGEMENT

SCM digital transformation has been dramatically accelerated during 2021–2025 with the fast development of technologies, such as AI, IoT, blockchain, and cloud computing. Dynamic predictive learning: AI has made significant contributions toward better predictions, predictions of demand, and autonomous actions. Organizations can use it to process huge amount of data to have optimized inventory, predict disruptions, and increase operational flexibility (Hasan & Islam, 2024). AI-enabled tools, like those used in systems developed by optimal dynamics, are illustrative of how logistics processes are made agile in real time by intelligent automation (Financial Times, 2025). To the same extent, the IoT technology became an integral part of digital SCM by delivering timely data from the connected sensors and devices. These are technologies that provide detailed awareness of where the products are, in what conditions they are stored, in which their status of transportation, which is enabling the proactive management of risks and therefore improving customer satisfaction (IMC Certifications, n.d.; Unity SCM, n.d.).

Blockchain technology, in contrast, enables transparency, traceability, and security by generating unalterable records of transactions along the supply chain. This effect is most pronounced in industries in which compliance and authenticity verification processes are very strict, including the pharmaceuticals and food logistics sectors (Fauziah & Mashudi, 2023). Integration with a trusted blockchain model, meaning all parties involved will be working with a single source of truth, thereby minimizing fraud in the distributed environment increasing trust among the parties. Cloud computing acts like the digital skeleton for the emerging trends. It provides the speed with which real-time collaboration, data availability, and the speed and rapid scalability required to react to changing market conditions and international disruptive events (Maritime Fairtrade, 2023). Cloud platforms enable easy interoperability among supply chain partners, aid in data-driven decision making, and enable implementation of AI and IoT solutions on a standard and scalable IT infrastructure (Supply & Demand Chain Executive, 2025; Unity SCM, n.d.). Combined, these two technologies have not just disrupted operational efficacy but also have been instrumental for emergence of robust end-to-end agile and sustainable supply chains able to weather uncertainties in today's hyperconnected world.

6.3 THE ROLE OF DATA ANALYTICS IN ENHANCING SUPPLY CHAIN RESILIENCE

The importance of data analytics and particularly in predictive analytics, real-time monitoring, data-driven decision-making is constantly reported in the literature for the last five years on supply chain resilience (Table 6.1). Prediction has been acknowledged as a key enabler in predicting disruptions and maximizing operation continuity. Predictive models can combine historical data with external signals (for example, weather patterns, market performance, and geopolitical risk) to predict supply chain disruptions before they happen (Amos, 2024). This predictive capability has become instrumental for tactical inventory management and proactive risk management. Also, Ibadullaev (2023) emphasizes the visibility of monitoring in real time with the help of the integration of IoT sensors and data analytics platforms for the supply networks. This constant stream of real-time data enables organizations to quickly see and react to delays, quality problems, and inefficient routes, helping reduce downtime and increase agility.

In addition, decision-making in supply chain has advanced by the introduction of advanced analytical tools, such as digital twins and prescriptive analytics. KPMG (2024) points out that through decision-making, analytics enables businesses to model different scenarios, evaluate their effect on supply chain performance, and adopt suitable tactics even under uncertain conditions. Zheng and Brintrup (2025) further this discussion in investigating the potential of graph neural networks and federated learning in supply chains, to allow for decentralized data processing, and increased visibility and collaboration between partners. Such an approach facilitates more accurate privacy preserving decision-making in complex multilayer supply chain networks. Taken together, such studies confirm the importance of mastering AI and data science, monitoring and decision support, and AI decision support to engineer resilient, adaptive, and future-robust supply chains.

6.4 INTEGRATING SUSTAINABILITY GOALS INTO DIGITAL SUPPLY CHAIN STRATEGIES

Today, in the business sector, it is becoming increasingly necessary to incorporate ESG (Environmental, Social, and Governance (ESG) principles in digital supply chain strategies. This inclusion satisfies not only regulatory obligations but also supports stakeholder expectations and resilience building. The integration of digital applications with ESG goals enables transparency, efficiency, and ethical supply chains to materialize. The role of digital transformation in greening supply chains: What role does digital transformation have to play in advancing environmental sustainability of supply chains? AI, IoT, blockchain, and other technologies make real-time monitoring, predictive analytics, and traceability possible in order to mitigate any environmental impacts. AI-based analytics helps to optimize logistics, which in turn helps to reduce fuel consumption as well as decrease greenhouse gas emissions. For example, AI is able to accurately predict demand; this can result in an effective stock management and less wastage (Hasan, 2024). Sensors and fuel control technology and Advanced sensor technologies in airplanes help in predicting and controlling the fuel consumption and environmental friendliness of the aircraft. Under blockchain, news, transparency, and traceability can be achieved in the procurement of raw materials which may encourage sustainable procurement and may minimize the risk of abusing the environment. Social responsibility has no borders: How a software company in Brazil developed options for ethical business practices. Social responsibility: For centuries, overseas "Sweatshops" and child labor have been a source of cheap products that people in North America, Europe, and Japan have enjoyed. Fair work practices, work in the community, and supply chain human rights are the main concerns. Digital enables the monitoring and enforcement of social standards across the wide array of geographically distributed supplier base. They provide companies with the possibility to monitor labor conditions and to ensure that labor laws and ethical standards are respected (PwC, 2023). Blockchain and IoT can enable verifiable labor practices, thereby allowing the prevention of human rights abuses, and the promotion of ethical sourcing (Chen et al., 2024). Collaborative Digital Networks Enable Capacity Building Schemes to enable companies to assist suppliers in enhancing labor standards and social performance (Preprints.org, 2024).

Governance is the framework of systems and processes for holding corporations accountable, fair, and open in their dealings with stakeholders. By enabling better compliance management as well as more effective stakeholder engagement, digital transformation can contribute to better governance. Digital models simplify ESG reporting, facilitating timely and accurate actualization of sustainability performance, which is a prerequisite for compliance and invocation of investor confidence (e.g., Deloitte, 2023). Risk identification and management is supported by advanced analytics, which enables companies to address the potential governance problems in their supply chain proactively (MDPI, 2025). Stakeholder engagement is backed and enhanced by the availability and transparency of information, which in turn promotes trust and transparency (KPMG, 2023).

Including ESG values in digital supply chain strategies is critical in establishing resilient, sustainable, and ethical supply chains. These new digital tools and platforms serve as enablers, accelerating environmental stewardship, social responsibility, and robust governance. Companies that successfully harness these technologies will be best placed to address regulatory requirements, meet the expectations of their stakeholders, and achieve their long-term sustainability targets.

6.5 RISK MANAGEMENT AND CRISIS RESPONSE IN INTELLIGENT SUPPLY CHAINS

Supply chains of the future risk management/addressing the crisis risk management and crises response will become key for resilient, intelligent supply chains. COVID-19 acted as a wake-up call which highlighted weaknesses in the supply chain and drove companies to revisit supply chain resilience. This has been further illustrated by subsequent geopolitical conflicts such as the Russia-Ukraine war and Middle East tensions which emphasize the need for smart and adaptable supply chain systems. An important progress in this context is the maturity of the usage of AI and machine learning (ML) techniques to risk measurement in the supply chain. Jahin et al. (2023) reviewed the literature and pointed that AI/ML models such as Random Forest or XGBoost allowed to improve prediction, thus identifying better risk mitigation strategies. These tools allow nearly instantaneous access to large systems, helping respond faster to emerging issues. The idea of digital twins has also been popularized as a means of detecting and responding to disruptions. Ashraf et al. (2023) proposed a deep learning-based hybrid-cognitive digital supply chain twin framework, deep autoencoder neural networks, and long short-term memory models to detect the disruptions and predict the recovery times. Such a strategy can achieve dynamic simulation and online measurement during a crisis, which is helpful to the decision-making in the crisis. Agent-based simulation models have been used to assess supply chain resilience strategies in pandemics. Lazebnik (2025) developed a new model, which combined epidemiological dynamics and economics, to study the impact of different resilience options. The overall lesson for resilience planning is that a prudent strategy works better under both pandemic and non-pandemic conditions, and that preparedness and flexibility matter.

Global supply chains have been highly affected by geopolitical tensions due to which strategic changes have become indispensable. Pugu and Dzauharoh (2025), for example, investigated the effect of geopolitical instability on supply chain resilience and recommended that managerial practices for risk management such as securing supplier diversity and adoption of advanced technologies be incorporated into the supply chain. Marsh's Political Risk Report (2025) also emphasized that growing protectionism and changing trading regulations created unstable environments and that companies should put in place robust contingency plans to manage these emerging risks. There are empirical evidence on risk management and effective utilization. For example, one study by the International Journal of Logistics Management (2021) analyzed how logistics service providers adjusted to the pandemic by exploring agility, robustness, and learning from experience as an essential aspect of resiliency. These results emphasize the importance of dynamic capabilities and learning a new in supply chain risk management. To summarize, the literature of 2021–2025 focuses on the role of AI, ML, and Digital Transformation in operationalizing the strength of supply chain operations. Indeed, a proactive risk management strategy based on empirical understanding, rigorous modeling and analysis is definitely necessary to tackle the intricacy of the present-day supply chain in the face of enduring disruptions and uncertainties.

6.6 CE AND GREEN INNOVATION IN SUPPLY CHAINS

The combination of CE, green innovation, and supply chain has been paid much attention in academia and industry. Focus areas are sustainable supply chain, waste prevention, and product lifecycle management to increase environmental sustainability and operational excellence. Sustainable sourcing has become the basis of circular supply chains. Studies have shown that incorporating circular economic practices into the supply chain requires cooperation at all levels of the supply chain such as suppliers, manufacturers, distributors, and customers. Closed-loop systems can be easily realized, whereby resources are reused at maximum and materials are cycled back to the system, improving transparency and traceability of resources by the use of new technologies such as IoT and blockchain (Abid et al., 2024). Waste minimization measures have also been a key area. For example, in Brazil, companies such as Veja collaborate with recycling cooperatives, also called Catadores, to upcycle plastic waste into sustainable sneakers, thereby illustrating the socio-economic and environmental merits related to them (Time, 2024). Likewise, the United Kingdom is initiating initiatives such as Project Reclaim and the Circular Fashion Innovation Network in order to create a national textile recycling industry that will be capable of processing large quantities of textile waste (Vogue Business, 2025). Digitalization has improved product lifecycle management. The concept of material passport—digitalized documentation containing all components of a product—has been developed to support the strategic design of circularity in supply chains. These passports facilitate the identification of value for recovery, recycling, and re-use, thus helping the industry in becoming more circular (Wikipedia, 2024a). Furthermore, AI and ML technologies have been critical to progress CE goals (Waqas et al., 2025). Artificial technologies have been use to optimize resource flows, to establish sustainability objectives, to establish key performance indicators, and to reinforce the decision-making process of the supply chain (Hasan, 2024). Furthermore, AI-based classification systems have increased the accuracy and efficiency of the textile sorting processes and the move toward sustainable waste management (Spyridis et al., 2024). Policy regimes have been important as well. The Anti-Waste and Circular Economy Law require products to be repairable and prohibits the destruction of unsold goods in an attempt to prevent waste and a more CE (Wikipedia, 2024b). In the same way, Extended Producer Responsibility (EPR) has been suggested to make producers responsible for the life cycle of their products and promote green products and the internalization of waste management costs (Wikipedia, 2024c). Organizations are also moving toward more resilient and environmentally sustainable supply chain models by sustaining sourcing, waste and recycling, product lifecycle management, technological adoption, and policy incentives.

6.7 COLLABORATION AND TRANSPARENCY THROUGH DIGITAL SUPPLY CHAIN PLATFORMS

The digital supply chain platform literature highlights an explicit move toward collaboration and transparency, with blockchain, cloud computing, and supplier relationship management (SRM) as main enablers. Adoption of blockchain technology has been widely acknowledged to promote data authenticity, traceability, and trust in supply networks. Hellani et al. (2021) had underscored that blockchain's decentralized nature allows terminologically immutable transactions at a level of integrity and trust that makes fraud less likely and inter-organizational trust possible. Expanding on this, Jovanovic et al. (2022) examined the scenario of Trade Lens, a blockchain-enabled logistics solution that was created by Maersk and IBM, and found increasing cargo visibility and effective stakeholders' coordination. Cloud integration, in particular, has also been a game changer by supporting real-time data exchange, on-demand scalability, and the centralized organization of supply chain data. Sahoo and Goswami (2024) have highlighted that cloud computing enables the firms to interconnect the redundant supply chain entities assisting in data symmetry and agility. Their research also found that cloud computing increases responsiveness in the face of disruptions by offering predictive analytics and automating coordination activities.

The use of digital platforms has greatly helped with more organized and embedded collaboration for Supplier Relationship Management (SRM). According to Goffin et al. (2022), these digital SRM tools allow organizations to maintain ongoing performance tracking with suppliers, risk management, and even transparency in objectives through common platforms to build long-lasting partnerships. Also, blockchain-enabled SRM systems, as proposed in Shan et al. (2023), provide a permanent record of supplier certifications and compliance, hold organizations accountable, and mitigate operational risk. Even in competitive sectors, we are seeing collaborative structures beginning to emerge, as the following examples illustrate. The Aura Blockchain Consortium, which includes LVMH, Prada, and Cartier, offered a compelling file-sharing case study of where blockchain could pull competing brands together toward shared transparency objectives, tracking product origins conscientiously still respecting brand sovereignty (Vogue Business, 2024). Whether or not those efforts are made, however, the industry-wide projects illustrate how powerful digital platforms are in helping to connect traditional silos. Despite these successes, there are challenges, specifically with respect to interoperability, and standardization of data across platforms. Vogue Business (2025) drew attention to an increasing frustration among supply chain leaders with a lack of commonality in digital reporting and sustainability metrics and the imperative to see harmonization around both frameworks and digital governance.

6.8 FUTURE TRENDS AND STRATEGIC ROADMAPS FOR RESILIENT DIGITAL SUPPLY CHAINS

Industry 5.0 is a direction which diverts from purely automated Industry 4.0 and toward collaboration where the focus is placed on the symbiotic relationship between humans and machines. This shift produces cyber-physical-social systems through incorporating social networks and cyber-physical systems, which will yield highly connected and dynamic industrial systems. This integrated approach is conducive to efficient resource management as well as logistics, improvisation, and resilience in supply chains (Lu et al., 2023). The introduction of so-called Collaborative Robots (COBOTS), which can easily switch tasks and cooperate with human workers, is an example of this trend.

Ethical AI is also being applied in SCM to tackle issues surrounding bias, transparency, and accountability. Large Language Models (LLMs) have been crucial in driving decision-making, prediction, and operational efficacy in the supply chain. By capitalizing on sophisticated data analytics and insights in real time, LLMs can accomplish better resource allocation, cost optimization, and faster response to market adjustments. Regarding the adoption of such solutions, the implementation of the technologies requires a sound ethical framework such as the avoidance of bias and privacy protection in order to ensure fairness and transparency in the AI procedures (Aghaei et al., 2025). AI governance frameworks are becoming more important for organizations to connect technological developments with societal and environmental values (Financial Times, 2024).

In light of global disruptions, agile global logistics systems have become an integral part of resilient supply chains. Such models encourage flexibility, regionalism, and the use of digital to manoeuver through the messiness of trade. By employing digital twins—simulations of the physical processes involved in the supply chain—organizations are able to model scenarios, recognize potential blockages, and form strategies to mitigate such blockages without directly impacting operations (Rebus, 2025). In addition, the use of AI-led tooling allows for on-the-minute risk evaluation and proactive analysis and decision-making to up the agility of supply chains to respond to unexpected issues (SAP Africa News Center, 2025). We would like to conclude that a transformative time frame in SCM, when human-centric technology and values become integrated and adaptive. These are changes that not only improve operational resilience but also bring supply chain practice into line with wider societal and environmental imperatives.

6.9 CONCLUSION

With increasingly complex and global supply chains, the melding of digital transformation and sustainable innovation is no longer a strategic choice but a goal for survival. This chapter has discussed the multi-faceted nature of creating a resilient supply chain that is also consistent with long-term ESG objectives. It started out with an exploration of resilience and sustainability in a digital age. We looked at technology drivers like AI, IoT, blockchain, and cloud computing changing how supply chains operate. These capabilities enable organizations to react quickly to market changes, drive efficiency, and predict and prevent potential disruptions. The importance of the use of data analytics was underscored as a key enabler of proactive decision-making and real-time agility, and ultimately the resilience of the supply chain. Incorporating sustainability within digital Planning, particularly via ESG frameworks, attenuates reputation and regulatory risks and builds stakeholder confidence and competitive strength for an organization. The lessons taken from recent global shocks such as the COVID-19 pandemic demonstrated the importance of managing risks and being prepared for crisis. At the same time, the increasing importance of CE models and green innovations emphasize that linear consumption is being replaced by regenerative value chains. Collaboration, transparency, and trust—enabled by digital platforms—emerged as important enabling factors for agile, ethical, and responsive networks. Going forward, the supply chain will be driven by both technological advancements and changes in regulation, with undercurrents pointing toward a society that wants businesses to be environmentally and socially responsible. In short, companies that seize digital transformation opportunities, integrate sustainability into their DNA, and create resilient, data-powered, and ethically attuned supply chains will stand best placed to succeed in an uncertain and increasingly dynamic global economy.

REFERENCES

Abid, I., Fuad, S. M. Z. A., Chowdhury, M. J. M., Chowdhury, M. S. S., & Ferdous, M. S. (2024). A systematic literature review on the use of blockchain technology in transition to a circular economy. arXiv preprint arXiv:2408.11664. https://arxiv.org/abs/2408.11664

Aghaei, R., Kiaei, A. A., Boush, M., Vahidi, J., Barzegar, Z., & Rofoosheh, M. (2025). The potential of large language models in supply chain management: Advancing decision-making, efficiency, and innovation. arXiv preprint arXiv:2501.15411. https://arxiv.org/abs/2501.15411

Amos, Z. (2024, July 23). How predictive analytics helps mitigate supply chain disruptions. Data Science Central. https://www.datasciencecentral.com/how-predictive-analytics-helps-mitigate-supply-chain-disruptions/

Ashraf, M., Eltawil, A., & Ali, I. (2023). Disruption detection for a cognitive digital supply chain twin using hybrid deep learning. arXiv preprint arXiv:2309.14557. https://arxiv.org/abs/2309.14557

Bag, S., Wood, L. C., Xu, L., & Dhamija, P. (2023). Digital transformation and supply chain sustainability: A systematic literature review and future research agenda. Technological Forecasting and Social Change, 187, 122282. https://doi.org/10.1016/j.techfore.2022.122282

Chen, Y., Zhang, X., & Liu, H. (2024). Integrating ESG Principles into Smart Logistics: Toward Sustainable Supply Chains. Preprints.org. https://www.preprints.org/manuscript/202411.1432/v1

Choi, T. M., Guo, S., & Luo, S. (2021). Risk analysis in logistics systems: A research agenda during and after the COVID-19 pandemic. Transportation Research Part E: Logistics and Transportation Review, 145, 102190. https://doi.org/10.1016/j.tre.2020.102190

Field, A. (2023). ESG Disclosure Challenges? A Digital Plan Can Help—and Open New Routes to Profit. The Wall Street Journal. https://deloitte.wsj.com/sustainable-business/esg-disclosure-challenges-a-digital-plan-can-helpand-open-new-routes-to-profit-fb485f1e

Mashudi. & Fauziah, L., (2023). Digital transformation in supply chain management: The impact of blockchain technology implementation on operational efficiency. Branding: Jurnal Manajemen dan Bisnis, 3(2).

Murray, S. (2024, May 1). What does AI mean for a responsible business? Financial Times https://www.ft.com/content/52249269-cca7-4060-8009-ea1c1fa28f60

Wright, M. (2025, March 11). Companies seek AI solutions to supply chain fragility. Financial Times. https://www.ft.com/content/1d07a823-43da-4c1b-84d3-7e453ebb1b16

Goffin, K., Seuring, S., & Müller, M. (2022). Understanding the impact of digital transformation on supply chain collaboration. ResearchGate. https://www.researchgate.net/publication/381310444_Understanding_the_Impact_of_Digital_Transformation_on_Supply_Chain_Collaboration

Hasan, M., & Islam, M. (2024). AI-driven IoT and blockchain integration in Industry 5.0: A systematic review of supply chain transformation. Innovatech Engineering Journal, 99–116. https://doi.org/10.70937/itej.v1i01.12

Hasan, M. R. (2024). AI-enhanced decision-making for sustainable supply chains: Reducing carbon footprints in the USA. arXiv. https://arxiv.org/abs/2501.10364

Hellani, H., Sliman, L., Samhat, A. E, & Exposito, E. (2021). On blockchain integration with supply chain: Overview on data transparency. Logistics, 5(3), 46. https://doi.org/10.3390/logistics5030046

Ibadullaev, N. (2023). Real-time supply chain visibility: Leveraging IoT and data analytics for real-time monitoring and insights. Towards AI. https://pub.towardsai.net/real-time-supply-chain-visibility-leveraging-iot-and-data-analytics-for-real-time-monitoring-and-bb156c42140f

IMC Certifications. (n.d.). Digital transformation in supply chain management for the logistics and transportation sector. https://imcinstitute.ae/digital-transformation-in-supply-chain-management-for-the-logistics-and-transportation-sector

International Journal of Logistics Management. (2021). Supply chain risk management in the COVID-19 pandemic: Strategies and empirical lessons for improving global logistics service providers' performance. https://www.emerald.com/insight/content/doi/10.1108/ijlm-02-2021-0109/full/html

Ivanov, D., & Dolgui, A. (2022). A digital supply chain twin for managing the disruption risks and resilience in the era of Industry 4.0. Production Planning & Control, 33(2–3), 235–251. https://doi.org/10.1080/09537287.2021.1882693

Jahin, M. A., Naife, S. A., Saha, A. K., & Mridha, M. F. (2023). AI in supply chain risk assessment: A systematic literature review and bibliometric analysis. arXiv preprint arXiv:2401.10895. https://arxiv.org/abs/2401.10895

Jovanovic, M., Kostić, N., Sebastian, I. M., & Sedej, T. (2022). Managing a blockchain-based platform ecosystem for industry-wide adoption: The case of Trade Lens. arXiv preprint arXiv:2209.04206. https://arxiv.org/abs/2209.04206

KPMG. (2023). Future of supply chain: ESG, technologies, workforce. https://kpmg.com/au/en/home/insights/2023/09/future-supply-chain-esg-technologies-workforce.html

KPMG. (2024, October). Strategic role of data and analytics in modern supply chains. https://kpmg.com/in/en/blogs/2024/10/strategic-role-of-data-and-analytics-in-modern-supply-chains.html

Kumar, S., Raut, R. D., Govindan, K., Mangla, S. K., & Yadav, G. (2024). Blockchain technology integration with sustainability and resilience in supply chains: A systematic review and future directions. Journal of Cleaner Production, 418, 139859. https://doi.org/10.1016/j.jclepro.2023.139859

Lu, Y., Xu, X., & Wang, L. (2023). Towards supply chain 5.0: Redesigning supply chains as resilient, sustainable, and human-centric systems in a post-pandemic world. Operations Research Forum, 4(1), 1–20. https://link.springer.com/article/10.1007/s43069-023-00234-3

Lazebnik, T. (2025). Evaluating supply chain resilience during pandemic using agent-based simulation. Physica A: Statistical Mechanics and its Applications, 674, Article 130780. https://doi.org/10.1016/j.physa.2025.130780

Maritime Fairtrade. (2023). Impact of digital transformation on supply chain technology. https://maritimefairtrade.org/impact-of-digital-transformation-on-supply-chain-technology/

Marsh. (2025, March 10). Global geopolitical tensions and increased protectionism destabilizing supply chains in 2025. https://www.marsh.com/en/about/media/global-geopolitical-tensions-and-increased-protectionism-destabilizing-supply-chains-in-2025.html

MDPI (2025). The impact of digital transformation on ESG performance: The role of supply chain resilience. Sustainability, 16(17), 7621. https://www.mdpi.com/2071-1050/16/17/7621

Preprints.org. (2024). Sustainability practices in supply chain management: Insights from leading global firms. https://www.preprints.org/manuscript/202406.0926/v1

Pugu, M. R., & Dzauharoh, S. (2025). The impact of geopolitical uncertainty on global supply chain resilience. International Journal of Society Reviews, 3(2), 360–366.

PwC. (2023). Supply chain sustainability. https://www.pwc.com/us/en/services/consulting/business-transformation/digital-supply-chain-survey/sustainability-in-supply-chains.html

Rebus. (2025). Supply chain trends for 2025: 7 key trends to the future. https://rebus.io/blog/supply-chain-trends-for-2025/Rebus+1InnovateXLabs+1

Sahoo, S., & Goswami, S. (2024). Cloud computing in supply chain management: Exploring the relationship. ResearchGate. https://www.researchgate.net/publication/370501433_Cloud_computing_in_supply_chain_management_Exploring_the_relationship

SAP Africa News Center. (2025, February 1). AI, sustainability and resilience top supply chain trends for 2025. https://news.sap.com/africa/2025/02/ai-sustainability-and-resilience-top-supply-chain-trends-for-2025

Shan, Y., Wang, W., & Song, M. (2023). Blockchain-enhanced supplier management and its impact on supply chain performance. Management Science, 69(4), 2048–2064. https://doi.org/10.1287/mnsc.2023.4851

Spyridis, Y., Argyriou, V., Sarigiannidis, A., Radoglou, P., & Sarigiannidis, P. (2024). Autonomous AI-enabled industrial sorting pipeline for advanced textile recycling. arXiv preprint arXiv:2405.10696. https://arxiv.org/abs/2405.10696

Supply & Demand Chain Executive. (2025). Supply chain technology in 2025: Advancements, benefits and challenges ahead. https://www.sdcexec.com/software-technology/emerging-technologies/article/22927352/supply-chain-technology-in-2025-advancements-benefits-and-challenges-ahead

Time. (2024, December 15). How Brazil recycling co-ops are helping turn plastic waste into shoes. https://time.com/7198461/brazil-recycling-co-ops-helping-turn-plastic-waste-into-shoes-veja/

Unity SCM. (n.d.). Embracing digital transformation in supply chain management. https://www.unityscm.com/blog/embracing-digital-transformation-in-supply-chain-management

Vogue Business. (2024c, June 15). How the Aura Blockchain Consortium convinced luxury competitors to collaborate. https://www.voguebusiness.com/story/technology/how-the-aura-blockchain-consortium-convinced-luxury-competitors-to-collaborate

Vogue Business. (2025a, February 17). Confused by supply chain reporting rules? You're not the only one. https://www.voguebusiness.com/story/sustainability/confused-by-supply-chain-reporting-rules-youre-not-the-only-one

Vogue Business. (2025b, May 9). Can the UK become a global leader in circular textiles? https://www.voguebusiness.com/story/sustainability/can-the-uk-become-a-global-leader-in-circular-textiles

Waqas, U., Umair, S., Mrugalska, B., & Ali Sulaiman, M. A. B. (2025). Twin transformation: The role of digitalization with artificial intelligence, big data and circular supply chain in circular economy performance for sustainability. Equilibrium. Quarterly Journal of Economics and Economic Policy, 20(2), 613–645. https://doi.org/10.24136/eq.3350.

Wikipedia contributors. (2024a, February 10). Material passport. Wikipedia. https://en.wikipedia.org/wiki/Material_passport

Wikipedia contributors. (2024b, February 10). Anti-waste and circular economy law. Wikipedia. https://en.wikipedia.org/wiki/Anti-Waste_and_Circular_Economy_Law

Wikipedia contributors. (2024c, February 10). Extended producer responsibility. Wikipedia. https://en.wikipedia.org/wiki/Extended_producer_responsibility

Zhang, Y., Li, Y., & Wang, J. (2025). Building sustainable and resilient supply chains in the circular economy: A multi-method approach. Journal of Business Research, 168, 114175.

Zheng, G., & Brintrup, A. (2025). An analytics-driven approach to enhancing supply chain visibility with graph neural networks and federated learning. arXiv preprint arXiv:2503.07231. https://arxiv.org/abs/2503.07231

7 Commitment of Logistics Companies to Sustainability Agenda

Comparative Study of GCC and ASEAN Countries

Ilya Bystrov, Mohamed Wahish Hashim, Sonia Umair, and Umair Waqas

7.1 INTRODUCTION

The contemporary world is rich with new phenomena that were not existent just half a century ago but which today form everyday reality through a system of complex interrelations. Globalization (Abeysekera 2022; Bystrov, Hashim and Kumar 2024; Hallin, Karrbom-Gustavsson and Dobers 2021), the Industrial Revolution 4.0 (Anh et al. 2022; Umair et al. 2024), the circular economy (Medaglia, Rukanova and Zhang 2024), sustainability (Fry and Egel 2021; Waqas, Umair and Mrugalska 2024), digitalization (Radicic and Petković 2023; Tagscherer and Carbon 2023), artificial intelligence (AI) (Holmström 2022; Jonathan, Magd and Khan 2024), and many other concepts and phenomena are reshaping the way human beings live in society, as well as how organizations approach their everyday operations and their long-term aspirations to success.

The two central phenomena that form the key theme of this book are digitalization and sustainability. We would like to highlight that, as the following section of this chapter demonstrates, many studies establish a strong connection between them. In addition, digitalization is an approach that companies worldwide have applied in all sectors of the economy. Although this approach was first employed to increase overall efficiency, achieve better financial results, and guarantee the long-term prosperity of businesses, as the sustainability idea gained support around the world, digitalization emerged as a crucial weapon in its toolbox. If we examine logistics organizations, we will find that nearly all of them have some aspect of their business processes digitalized; the only difference is the degree of digitalization. However, the extent to which such widely applied digitalization contributed to the commitment of logistics companies to sustainability requires evaluation.

This chapter's goal is to gauge how committed logistics businesses are to the sustainability agenda in the two Asian and Middle Eastern associations' member

DOI: 10.1201/9781003611905-10

nations. After reading this chapter, the reader will be able to compare the level of commitment to sustainability principles of logistics organizations in the Gulf Cooperation Council (GCC) and Association of Southeast Asian Nations (ASEAN) and explain how sustainability and digitalization concepts are related.

The structure of this chapter is as follows: Introduction, Literature Review, Methodology, Results and Discussion, and Conclusions and Recommendations. The introduction section provides the background description and the learning and research goals. The link between sustainability and digitalization is reviewed in the literature review section, which also includes a brief discussion of contemporary research that concentrates on how businesses in general and logistics companies in particular can benefit from following sustainable practices. While the results and discussion section includes the study's findings and an interpretation of them, the methodology section outlines the strategy we employed to achieve the research goal. The final portion of the study, the conclusions and recommendations, highlights the main conclusions and provides fundamental suggestions to company-level decision-makers.

7.2 LITERATURE REVIEW

7.2.1 Relation between Sustainability and Digitalization

There are numerous connections between sustainability and digitalization. The main effect of digital technologies comes from the way they function: they make things more efficient, allow environmentally friendly methods, and make it easier for companies and industries to use sustainable practices. Below, we discuss some of the major contributions that digitalization impacts.

The first major impact that digitalization has on sustainability is the improved capabilities of resource usage optimization. Such digital tools, like the Internet of Things and AI, can dramatically change the way the usage monitoring and optimization of key resources like energy, water, or raw materials are done (Hasan et al. 2024; Lorenc and Kuźnar 2024). The immediate consequence of this would be a significant reduction in waste and, in this way, promotion of sustainability. Multiple publications concentrate on this (Waqas, Umair and Mrugalska 2024). For instance, a recent research (Assareh et al. 2024) focuses on the application of AI in optimizing the water desalination process, which utilizes the waste heat from a thermal power plant. The suggested method allows for decreasing carbon emissions, as well as increasing power output and efficiency. An earlier explorative, mixed-method study (Bang and Andersen 2022) explored the potential applications of AI in mitigating waste production on construction sites. The study draws conclusions about AI-based tools and instruments that can enhance waste management in construction projects.

The next aspect where digitalization makes its positive impact is the development of the smart city concept. By strengthening public transportation, improving urban planning, and lowering environmental footprints through interconnected infrastructure, digitalization contributes to the promotion of this concept (Goh et al. 2024; Kong et al. 2024). If we go deeper into the details, a recent study (Wang, Xu and Wu 2024) based on the empirical data from China investigates the role of the Open

Government Data, which is enabled by digitalization, in smart city development. In a similar vein, another study (Zhou and Liu 2024) reviews the development status and mechanisms of energy digitalization technologies for high-efficiency, low-carbon, and intelligent building energy systems. For adapting to climate change and making transitions to smart cities, the research directly emphasizes the importance of digitalization technologies for energy flexibility and resilience.

Digitalization also positively contributes to advancing circular economy principles, which is important, as circular economy is perceived as instrumentalizing and operationalizing sustainable development (Ghisellini, Cialani and Ulgiati 2016; Murray, Skene and Haynes 2017), which some authors believe is vague to some extent (Hickel 2019; Ward et al. 2016). Such technologies as blockchain and the Internet of Things enable better capabilities in tracking and recycling materials (Borandag 2023; Gong et al. 2022), which promote the circular economy by reducing the usage of virgin resources and reusing them. Another study (Hong and Xiao 2024) explores the potential of blockchain and AI in supply chain coordination and mitigation of negative impacts. It says that the technologies listed above can encourage usage of recycling and circular business models. Additionally, they can simplify tasks such as carbon accounting, offsetting, and other activities that align with the circular economy's principles. The study also offers policy recommendations for countries to adopt smart and green industrial transformation pathways. According to the study, using these technologies and methods would help governments and other policymakers make the taken actions and activities more transparent, as well as more traceable. Such changes could further reduce environmental threats, improve production systems, and support achievement of sustainability goals in various other ways. Another recent study (Dwivedi, Ganguly and Paul 2024) focused on critical success factors for the adoption of digital technologies in circular supply chains. The first five critical success factors that were highlighted in the research included employers' institutional commitment, the use of current technologies, the endorsement of the organizational chief officer, current forms of business models, and modernization that relies on available data.

The standard solutions that are enabled by digitalization make it possible to ensure better energy management, promote renewable energy integration, and reduce emissions (He et al. 2025; Li, Zhao and Wang 2024; Y. Zhao and Wang 2025). In this way, a recent study (H. Zhang, Zhang and Pai 2024) based on the empirical data covering 32 provinces of China in the period between 2010 and 2021, which is dedicated to the promotion of carbon neutrality and green growth, calls for advancing digitalization as a possible strategy. Another study (Mehta and Shah 2024) that addresses carbon emissions in BRICS countries reconfirms the transformative effect of digitalization and its role in the reduction of carbon emissions.

Digitalization contributes to the development of every industry one can think of—finance (Zhao and Xing 2024), construction (Fonseca, Benito and Piña Ramírez 2024), or education (Yonghua Zou et al. 2024). For instance, digitalization in agriculture, which is realized through the application of drones, sensors, and AI, leads to improved yields and optimization of resource usage (Khan et al. 2024; Khodjaev et al. 2025; Thai and Le 2025). Overall, just a brief look at publications related to digitalization in agriculture allows one to understand that this topic is addressed by scholars from all across the world—China (Zhang et al. 2024), Ghana (Abdulai et al. 2023), or Russia (Begishev et al. 2024).

In the following section of the chapter, we discuss the impact that the implementation of sustainable practices in companies causes. It's not surprising that an attentive reader will immediately recognize the application of instruments related to digitalization behind these practices.

7.2.2 Effects of Implementing Sustainability in Business Practices

The Internet of Things and AI, two tools for digitalization we talked about above, make operations more efficient. Numerous studies in a variety of fields, including healthcare (Kinachtchouk and Canes 2025), education (Kurata et al. 2025), and agriculture (Elhattab and Elatar 2025), have demonstrated this. The same technologies applied in logistics companies and supply chains lead to reduction of idle time and more efficient fleet management (Merlo 2024), decrease in fuel consumption (Whig et al. 2024), minimization of packaging waste (Abekoon et al. 2024), and other effects that not only lower operational costs but substantially contribute to the company's advancement on the way to sustainability.

Regulations pertaining to the application of sustainability principles vary across different regions and nations. It is not a secret that more developed countries have advanced comparatively further on the road of the introduction of sustainability-related regulations (Singhania and Saini 2023). Adhering to environmental regulations in such countries helps companies to avoid fines and legal complications; in simple words, it is not a matter of gaining a competitive advantage, but rather a prerequisite for survival. Laws and rules usually reflect the most important values of the area or country.

Because of this, companies in general, and logistics companies in particular, that work in more developed countries should know that customers usually expect them to take actions and make decisions that are beneficial for the environment. This applies to most industries, be it tourism (Tasci 2017), food manufacturing (Piracci, Boncinelli and Casini 2022), or logistics and organization of supply chains (Pérez-Mesa et al. 2019; Shen and Panichakarn 2023; Ting et al. 2014). Companies that manage to fulfill this requirement enjoy improved reputation (Díaz-Cruces, Méndez Rocasolano and Miró Colmenárez 2024; Hao et al. 2025; Yuxuan Zou et al. 2025) and higher brand value (Bhat, Gupta and Singh 2024; Jagani, Saboorideilami and Tarannum 2024; Lee, Lee and Park 2024).

Worth mentioning, the positive impact extends beyond the external environment. Multiple publications report that company employees can positively perceive the company's efforts to minimize the environmental impact (Ahmad, Hidthiir and Rahman 2024; Ahmed et al. 2024; Nagiah and Mohd Suki 2024). If we go into more details, a recent study (Obrecht, Ojsteršek and Pavić 2024) conducted among 596 employees representing different industries stipulates that people that are following patterns of sustainable behavior prefer to work in companies that implement sustainable practices in their operations. Another study (Alherimi et al. 2024) among employees of a UAE-based automotive company reported similar results, demonstrating that eco-friendly practices attract environmentally conscious employees.

7.3 METHODOLOGY

In order to fulfill the study's research objective, we decided to use content analysis method. For our comparative investigation, we chose ten companies from each of the GCC and ASEAN nations. For our sample, we chose only those companies that were established at least five years ago, as in this case, we could be sure that they have already survived their infancy stage and started to formalize their activities and strategy. Thus, the sample size for the ASEAN region was 100 enterprises, and the sample size for the GCC region was 60 companies. We did not differentiate between the different types of logistics services that the companies were rendering, nor did we attempt to have a balanced representation of companies specializing in transportation, shipping, warehousing and distribution, or freight forwarding, as we observed that most logistics companies in the studied regions combine multiple logistics-related services in their activities. Consequently, we chose 160 companies for our analysis, dividing them as Table 7.1 shows.

In a similar manner as in our earlier studies (Bystrov and Panigrahi 2021; Magd and Bystrov 2021), we carried out the analysis in two stages:

Step 1: We looked for terms and phrases in mission and vision statements that would show the commitment of the 160 sampled companies to the sustainability agenda.

Step 2: We mapped the content of secondary strategic documents, such as objectives, values, and similar, to the Sustainable Development Goals (SDGs).

TABLE 7.1
Summary of the Study Sample

Association/Country	Number of Selected Companies
ASEAN, including	**100**
Brunei	10
Cambodia	10
Indonesia	10
Laos	10
Myanmar	10
Malaysia	10
Philippines	10
Singapore	10
Thailand	10
Vietnam	10
GCC, including	**60**
Bahrain	10
Kuwait	10
Oman	10
Qatar	10
Saudi Arabia	10
United Arab Emirates	10

The most prevalent terms that link these statements to the SDGs were also recorded.

The analysis concludes with a comparison of the results demonstrated by the companies that belonged to the studied regional associations.

7.4 RESULTS AND DISCUSSION

The majority of the GCC logistics companies under study made their most important strategic documents available to the public. We found that 48 of the 60 companies that were chosen, or 80% of them, have vision and mission statements, goals, and/or values on their websites. In a similar vein, 81 out of 100 chosen organizations, or 81%, are ASEAN logistics companies that make these statements publicly available. Access to the necessary documentation was greatly facilitated by the bilingual websites of nearly all of the chosen businesses.

In the first stage of the analysis, we measured logistics organizations' overall commitment to the sustainability agenda by identifying terms like "sustainability," "sustainable," "sustainable development," or similar in mission and vision statements. Both regions demonstrated very moderate results. It should be noted that some sampled companies did not make a direct reference to sustainability in their key strategic statements but had a section dedicated to it on their websites. It could be either a section addressing sustainability directly or via environmental, social, and governance (ESG) reporting. For the purposes of our research, we considered such sections the equivalent of direct referencing in the strategic statements. In the study, 23 out of 160 companies, or 14.5%, had such a reference, based on the average for the both areas. While GCC companies showed slightly below-average commitment with 11.5% (or 7 out of 60 sampled companies), ASEAN logistics companies displayed slightly above-average results, with 16% (16 out of 100) companies mentioning sustainability in their mission and vision statements, or having a special section dedicated to sustainability on the website. Still, the results for both regional associations were very close. One could name these cases as demonstrating a firm commitment to the sustainability agenda.

After analysis of the mission and vision statements, we moved on to the next step, which was to take a closer look at the logistics companies' goals or objectives, values, and similar second-level strategic documents. The followed procedure supposed mapping of the contents of these documents to 17 SDGs. In a similar way, we mapped the contents of specialized areas of the companies' websites that were dedicated to sustainability, if they were available. Table 7.2 displays the results of this mapping.

Table 7.2 allows making several observations. First of all, logistics companies of both associations completely bypass SDG1, SDG2, and SDG4, which are related to accessibility to education and tackling poverty and hunger. In addition, SDG16 and SDG17, which focus on institutions, peace, and justice, as well as the establishment of partnerships, are not addressed in the strategic documents of logistics companies operating in ASEAN. Logistics companies operating in the GCC also address these SDGs minimally.

TABLE 7.2
Mapping of Strategic-Level Statements to Sustainable Development Goals

Sustainable Development Goal	ASEAN	GCC
SDG 1: No Poverty	No alignment	No alignment
SDG 2: No Hunger	No alignment	No alignment
SDG 3: Good Health and Well-Being	2 cases (2%)	3 cases (5%)
SDG 4: Quality Education	No alignment	No alignment
SDG 5: Gender Equality	5 cases (5%)	3 cases (5%)
SDG 6: Clean Water and Sanitation	12 cases (12%)	7 cases (11.5%)
SDG 7: Affordable and Clean Energy	13 cases (13%)	7 cases (11.5%)
SDG 8: Decent Work and Economic Growth	10 cases (10%)	6 cases (10%)
SDG 9: Industry, Innovation, and Infrastructure	17 cases (17%)	4 cases (6.5%)
SDG 10: Reduced Inequalities	5 cases (5%)	4 cases (6.5%)
SDG 11: Sustainable Cities and Communities	9 cases (9%)	2 cases (3.5%)
SDG 12: Responsible Consumption and Production	24 cases (24%)	6 cases (10%)
SDG 13: Climate Action	20 cases (20%)	6 cases (10%)
SDG 14: Life below Water	13 cases (13%)	3 cases (5%)
SDG 15: Life on Land	11 cases (11%)	3 cases (5%)
SDG 16: Peace, Justice, and Strong Institutions	No alignment	1 case (1.5%)
SDG 17: Partnerships for the Goals	No alignment	2 cases (3.5%)

Overall, companies of the GCC region don't have "favorite" SDGs with which their strategic statements are aligned the most. There is a group of "preferred" SDGs that are only marginally ahead of others—SDG6–8 and SDG12–13. Mapping companies of the ASEAN region gave a different picture. SDG12 and SDG13, which pertain to responsible consumption and production as well as climate action, stand out as absolute leaders, followed by a tight group of SDG6–8 and SDG14–15, with SDG9 taking the place in between the two groups.

Another interesting observation that our sample allows us to make is related to the countries with a relatively bigger number of logistics companies whose strategic statements are aligned with at least one SDG or that have a section dedicated to sustainability that are aligned in the similar way. The latter automatically means alignment with multiple SDGs. Table 7.3 provides a summary of this data.

As Table 7.3 shows, ASEAN has countries that could be called leading and countries that are lagging behind. Singapore, Thailand, and Vietnam belong to the lagging category, while Cambodia, Indonesia, and the Philippines occupy a relatively advanced position. Obviously, our samples were more oriented on the regional associations as a whole and they don't allow us to make a statistically correct conclusion; however, still, this gives an approximate evaluation and calls for further research. It is worth noting that a relatively bigger share of logistics companies operating in the ASEAN region—33%, compared to 18.5% in the GCC region—addresses at least one SDG.

Overall, we can evaluate the current state of commitment of logistics companies in the studied regions to the sustainability agenda as satisfactory. However, it needs

TABLE 7.3
Number of Logistics Companies per Country with Alignment to at Least One SDG

Country	Number of Logistics Companies with Positive Mapping to at Least One SDG
ASEAN	**33**
Brunei	3
Cambodia	5
Indonesia	6
Laos	3
Myanmar	4
Malaysia	3
Philippines	5
Singapore	1
Thailand	1
Vietnam	2
GCC	**11**
Bahrain	2
Kuwait	1
Oman	3
Qatar	1
Saudi Arabia	3
United Arab Emirates	1

to be noted that a significant contribution to the numbers discussed above is made by large logistics companies operating on a global scale and having a presence in the investigated regions. These businesses can show they take a more advanced approach to sustainability by, among other things, using ESG and/or sustainability reporting and including sustainability in their core values and goals. Logistics companies that are operating locally are definitely lagging behind their international counterparts.

More interesting conclusions can be drawn by future research that will focus on comparative analysis with regions that are considered leaders in this sense (e.g., the European Union) or attempt to evaluate the dynamics of changes with time. Further, we match and compare our study's observations with those from related research.

One of the first observations that we made was about the availability of key strategic documents on the companies' websites. Even though the level of 80% availability in the case of logistics companies from both studied regions is solid, it is still lower than in the case of, for instance, higher education institutions, where almost every studied entity (more than 90%) in the same regions had them and made them open to the public (Bystrov and Panigrahi 2021).

Logistics companies share a similar level of commitment to sustainability with higher education institutions. The share of companies and institutions that explicitly reference sustainability in their mission and vision statements or have a site section for it is almost the same in the current and previous (Magd and Bystrov 2021) research.

Another observation we made about the role of ESG reporting in certain countries and regions aligns with the findings of other studies. In this way, recent research (Imperiale, Pizzi and Lippolis 2023; Maalouf 2024; Paridhi and Arora 2023) shows that policymakers play a big part in putting sustainability-related reporting through ESG into action, but it also says that it's not always easy, accurate, or fault-free.

The overall progress of the studied regions on the way to sustainability is evaluated controversially by researchers. Some studies praise the achievements of ASEAN member countries on the way to sustainability (Damayanti, Sudarmaji and Masrio 2024), while even more publications criticize the mistakes and reveal challenges (Feng and Li 2024; Teow et al. 2024; Xu, Xu and Jia 2024). Similar observations can be offered regarding the GCC's achievements on its path to sustainability, with some positive evaluations (Alharbi and Csala 2020) and a greater number of publications that focus on existing challenges (Anwana and Owojori 2024; Nassar 2024; Umar and Umeokafor 2022). This aligns perfectly with our observations—a lot has been done, but even more still needs to be accomplished.

7.5 CONCLUSIONS AND RECOMMENDATIONS

In closing, we would like to draw attention to the following major findings and observations from our study:

- In both regions, 80%–85% of logistics companies disclose their mission and vision statements, meaning one in five either doesn't have them or doesn't disclose them.
- Logistics companies of both regions under consideration demonstrate moderate commitment to the sustainability agenda, with a limited number of businesses (11.5% in GCC and 16% cases in ASEAN samples) referencing sustainability directly in the mission and vision statements or having separate sections dedicated to sustainability.
- Mapping of the extended range of strategic documents—mission and vision statements, goals, values, and sustainability reports—to 17 SDGs showed that SDG1, SDG2, and SDG4 are completely missed by logistics companies of both regional associations. Additionally, logistics companies in ASEAN completely ignore SDG16 and SDG17, while companies operating in the GCC only minimally address them.
- Analysis of strategic-level statements and sustainability-related reports showed that logistics companies operating in the ASEAN region are more aligned with SDG12 and SDG13 (24% and 20% of logistics companies are aligned with them). These SDGs pertain to responsible production and consumption, as well as climate-related actions.
- Logistics companies operating in the GCC region do not demonstrate such selective affinity. A group of SDGs that led in the mapping exercise included five goals—SDG6, SDG7, and SDG8 (clean water and energy and economic growth), as well as SDG12 and SDG13 that were leading in the case

of logistics companies operating in ASEAN. Approximately, 10%–11.5% of logistics companies operating in the GCC region demonstrated alignment with these SDGs.
- Proportionally, the big international logistics companies demonstrate higher commitment to the sustainability agenda than smaller local companies.

Having evaluated the commitment of logistics companies to the sustainability agenda across both studied regional associations as overall positive, but insufficient, we would like to propose the following recommendations:

- Mission and vision statements are important instruments in the arsenal of strategic management and public relations. If a company already has mission and vision statements, it is highly recommended to make them available to the public, or if not, to start implementing strategic management practices and develop them.
- Given that our results revealed significantly higher levels of commitment to sustainability among large international logistics companies, we recommend the local companies also consider following this practice. If it doesn't happen, a potential partner or client that is operating globally and that has already adopted the corresponding values can make a choice in favor of another company that has similar values and is guided by the same principles in its operations.
- Local logistics companies, if they aspire to global growth, need to reassess their commitment to the sustainability agenda. In the context of global involvement, they will interact with countries that prioritize sustainability issues. Non-compliance with the corresponding regulations in these countries will raise questions about the further development of such companies.

In-depth research on a country-specific analysis, the inclusion of additional regional associations, or a shift from the strategic to the operational level and case studies demonstrating how operational plans and daily operations demonstrate strategic alignment with the sustainability agenda could be the main areas of future research. An interesting alternative that would demonstrate development dynamics would be to do the study again using the same methods after several years.

REFERENCES

Abdulai, Abdul-Rahim, Philip Tetteh Quarshie, Emily Duncan, and Evan Fraser. 2023. "Is Agricultural Digitalization a Reality among Smallholder Farmers in Africa? Unpacking Farmers' Lived Realities of Engagement with Digital Tools and Services in Rural Northern Ghana." *Agriculture & Food Security* 12 (1): 11. https://doi.org/10.1186/s40066-023-00416-6.

Abekoon, Thilina, B. L. S. K. Buthpitiya, Hirushan Sajindra, E. R. J. Samarakoon, J. A. D. C. A. Jayakody, Komali Kantamaneni, and Upaka Rathnayake. 2024. "A Comprehensive Review to Evaluate the Synergy of Intelligent Food Packaging with Modern Food Technology and Artificial Intelligence Field." *Discover Sustainability* 5 (1): 160. https://doi.org/10.1007/s43621-024-00371-7.

Abeysekera, Indra. 2022. "A Framework for Sustainability Reporting." *Sustainability Accounting, Management and Policy Journal* 13 (6): 1386–1409. https://doi.org/10.1108/SAMPJ-08-2021-0316.

Ahmad, Zaki, Mohammad Helmi Bin Hidthiir, and Md Mahfujur Rahman. 2024. "Impact of CSR Disclosure on Profitability and Firm Performance of Malaysian Halal Food Companies." *Discover Sustainability* 5 (1): 18. https://doi.org/10.1007/s43621-024-00189-3.

Ahmed, Zeeshan, Mishal Khosa, Nhat Tan Nguyen, Abdulaziz Fahmi Omar Faqera, and Shafique Ur Rehman. 2024. "Green HRM Fostering Followers' Green Creativity: Unfolding the Role of Environmental Concern and Employee Engagement with Eco-Initiatives as a Serial Mediation." *Creativity and Innovation Management* 33 (4): 742–65. https://doi.org/10.1111/caim.12614.

Alharbi, Fahad Radhi, and Denes Csala. 2020. "GCC Countries' Renewable Energy Penetration and the Progress of Their Energy Sector Projects." *IEEE Access* 8: 211986–212002. https://doi.org/10.1109/ACCESS.2020.3039936.

Alherimi, Nadin, Zeki Marva, Khalid Hamarsheh, and Ayman Alzaatreh. 2024. "Employees' Pro-Environmental Behavior in an Organization: A Case Study in the UAE." *Scientific Reports* 14 (1): 15371. https://doi.org/10.1038/s41598-024-66047-4.

Anh, Tran Duc, Tran Duc Binh, Nguyen Duc Bao Long, Tran Van Ai, Khong Sin Tan, and Nguyen Thi Le Van. 2022. "Strategic Vision for the Implementation of the Industrial Revolution 4.0 in the Vietnamese Context." *International Journal of Technology* 13 (5): 958. https://doi.org/10.14716/ijtech.v13i5.5838.

Anwana, Emem, and Oluwatobi Mary Owojori. 2024. "Policy Options for Achieving Sustainable Development Goals (SDGs) in the MENA Region through the Effective Implementation of Multilateral Environmental Agreements." *Journal of Sustainable Development Law and Policy (The)* 15 (2): 86–113. https://doi.org/10.4314/jsdlp.v15i2.4.

Assareh, Ehsanolah, Mohammad Firoozzadeh, Mohammad Zoghi, Ali Zare, Yasaman Ghazi, and Ali Shahin-Banna. 2024. "Water Desalination Using Waste Heat Recovery of Thermal Power Plant in Tropical Climate; Optimization by AI." *Energy Conversion and Management: X* 24 (October): 100731. https://doi.org/10.1016/j.ecmx.2024.100731.

Bang, Sofie, and Bjørn Andersen. 2022, September. "Utilising Artificial Intelligence in Construction Site Waste Reduction." *Journal of Engineering, Project, and Production Management*. https://doi.org/10.32738/JEPPM-2022-0022.

Begishev, Ildar, Albina Shutova, Yuri Vasiliev, Diana Bersei, Olga Perepadya, and Kirill Dolgopolov. 2024. "Digital Modernization of Agriculture." *BIO Web of Conferences* 140: 03017. https://doi.org/10.1051/bioconf/202414003017.

Bhat, Ishfaq Hussain, Shilpi Gupta, and Satinder Singh. 2024. "The Influence of Sustainability Communication on Consumer Loyalty and Brand Reputation." In *Contemporary Studies in Economic and Financial Analysis*, edited by Ridhima Sharma, Indira Bhardwaj, Simon Grima, Timcy Sachdeva, Kiran Sood, and Ercan Özen, 1–15. Emerald Publishing Limited. https://doi.org/10.1108/S1569-37592024000113A001.

Borandag, Emin. 2023. "A Blockchain-Based Recycling Platform Using Image Processing, QR Codes, and IoT System." *Sustainability* 15 (7): 6116. https://doi.org/10.3390/su15076116.

Bystrov, Ilya, Mohamed Wahish Hashim, and Vijayesh Kumar. 2024. "International Faculty as a Means to Promote Internationalization of an Educational Institution: An Important Consideration for Global Accreditation from GCC Perspective." In *Advances in Higher Education and Professional Development*, edited by Arshi Naim, Alok Saklani, Shad Ahmad Khan, and Praveen Kumar Malik, 186–99. IGI Global. https://doi.org/10.4018/979-8-3693-1698-6.ch012.

Bystrov, Ilya, and Shrikant Panigrahi. 2021. "Commitment of Higher Education Institutions to Sustainability Agenda: Evidences from GCC Countries." *Journal of Contemporary Issues in Business and Government* 27 (02). https://doi.org/10.47750/cibg.2021.27.02.438.

Damayanti, Setia, Eka Sudarmaji, and Herlan Masrio. 2024. "The Critical Role of Energy Intensity in Decarbonizing ASEAN: Integrating Growth and Emissions Reductions." *International Journal of Energy Economics and Policy* 14 (3): 247–59. https://doi.org/10.32479/ijeep.15059.

Díaz-Cruces, Eliana, María Méndez Rocasolano, and Pablo Javier Miró Colmenárez. 2024. "La Sostenibilidad Corporativa: Un Pilar Estratégico Para El Desarrollo Sostenible y La Competitividad Empresarial." *European Public & Social Innovation Review* 10 (December): 1–18. https://doi.org/10.31637/epsir-2025-795.

Dwivedi, Ashish, Anirban Ganguly, and Sanjoy Kumar Paul. 2024. "Critical Success Factors for Linking Digital Technologies and Circular Supply Chains." *Business Strategy and the Environment* 33 (8): 8332–60. https://doi.org/10.1002/bse.3907.

Elhattab, Kamal, and Said Elatar. 2025. "Evaluating Low-Cost Internet of Things and Artificial Intelligence in Agriculture." *Indonesian Journal of Electrical Engineering and Computer Science* 37 (2): 968. https://doi.org/10.11591/ijeecs.v37.i2.pp968-975.

Feng, Haiyan, and Yan Li. 2024. "The Role of Fintech, Natural Resources, Environmental Taxes and Urbanization on Environmental Sustainability: Evidence from the Novel Panel Data Approaches." *Resources Policy* 92 (May): 104970. https://doi.org/10.1016/j.resourpol.2024.104970.

Fonseca, Salazar Santos, Patricia Aguilera Benito, and Carolina Piña Ramírez. 2024. "Digital Horizons in Construction: A Comprehensive System for Excellence in Project Management." *Buildings* 14 (7): 2228. https://doi.org/10.3390/buildings14072228.

Fry, Louis, and Eleftheria Egel. 2021. "Global Leadership for Sustainability." *Sustainability* 13 (11): 6360. https://doi.org/10.3390/su13116360.

Ghisellini, Patrizia, Catia Cialani, and Sergio Ulgiati. 2016. "A Review on Circular Economy: The Expected Transition to a Balanced Interplay of Environmental and Economic Systems." *Journal of Cleaner Production* 114 (February): 11–32. https://doi.org/10.1016/j.jclepro.2015.09.007.

Goh, Kai Chen, Tonni Agustiono Kurniawan, Nadzirah Zainordin, Ika Diyah Candra Arifah, Muhamad Azahar Abas, Md Asrul Nasid Masrom, Sulzakimin Mohamed, et al. 2024. "Expediting Decarbonization in Energy, Waste, and Water Sector through Digitalization in Sustainable Smart Cities (SSC): Case-Studies in Malaysia and China Based on Industry 5.0 Paradigm." *Sustainable Cities and Society* 117 (December): 105969. https://doi.org/10.1016/j.scs.2024.105969.

Gong, Yu, Shenghao Xie, Deepak Arunachalam, Jiang Duan, and Jianli Luo. 2022. "Blockchain-Based Recycling and Its Impact on Recycling Performance: A Network Theory Perspective." *Business Strategy and the Environment* 31 (8): 3717–41. https://doi.org/10.1002/bse.3028.

Hallin, Anette, Tina Karrbom-Gustavsson, and Peter Dobers. 2021. "Transition towards and of Sustainability—Understanding Sustainability as Performative." *Business Strategy and the Environment* 30 (4): 1948–57. https://doi.org/10.1002/bse.2726.

Hao, Pu, Samar S. Alharbi, Ahmed Imran Hunjra, and Shikuan Zhao. 2025. "How Do ESG Ratings Promote Digital Technology Innovation?" *International Review of Financial Analysis* 97 (January): 103886. https://doi.org/10.1016/j.irfa.2024.103886.

Hasan, M. Z., M. Z. Hussain, S. Umair, and U Waqas. (2024). "Understanding Artificial Intelligence in Supply Chain and Innovation Performance." In *Human Perspectives of Industry 4.0 Organizations: Reviewing Sustainable Performance.* https://doi.org/10.1201/9781032616810-7.

He, Dawei, Xiangzheng Deng, Yunxiao Gao, and Xinsheng Wang. 2025. "How Does Digitalization Affect Carbon Emissions in Animal Husbandry? A New Evidence from China." *Resources, Conservation and Recycling* 214 (March): 108040. https://doi.org/10.1016/j.resconrec.2024.108040.

Hickel, Jason. 2019. "The Contradiction of the Sustainable Development Goals: Growth versus Ecology on a Finite Planet." *Sustainable Development* 27 (5): 873–84. https://doi.org/10.1002/sd.1947.

Holmström, Jonny. 2022. "From AI to Digital Transformation: The AI Readiness Framework." *Business Horizons* 65 (3): 329–39. https://doi.org/10.1016/j.bushor.2021.03.006.

Hong, Zexin, and Kun Xiao. 2024. "Digital Economy Structuring for Sustainable Development: The Role of Blockchain and Artificial Intelligence in Improving Supply Chain and Reducing Negative Environmental Impacts." *Scientific Reports* 14 (1): 3912. https://doi.org/10.1038/s41598-024-53760-3.

Imperiale, Francesca, Simone Pizzi, and Stella Lippolis. 2023. "Sustainability Reporting and ESG Performance in the Utilities Sector." *Utilities Policy* 80 (February): 101468. https://doi.org/10.1016/j.jup.2022.101468.

Jagani, Sandeep, Vafa Saboorideilami, and Saraf Tarannum. 2024. "Shaping Brand Attitudes through Sustainability Practices: A TSR Approach." *Journal of Services Marketing* 38 (3): 369–82. https://doi.org/10.1108/JSM-03-2023-0083.

Jonathan, Henry, Hesham Magd, and Shad Ahmad Khan. 2024. "Artificial Intelligence and Augmented Reality: A Business Fortune to Sustainability in the Digital Age." In *Navigating the Digital Landscape*, edited by Nripendra Singh, Pooja Kansra, and S. L. Gupta, 85–105. Emerald Publishing Limited. https://doi.org/10.1108/978-1-83549-272-720241005.

Khan, Sunawar, Tehseen Mazhar, Tariq Shahzad, Muhammad Amir Khan, Sghaier Guizani, and Habib Hamam. 2024. "Future of Sustainable Farming: Exploring Opportunities and Overcoming Barriers in Drone-IoT Integration." *Discover Sustainability* 5 (1): 470. https://doi.org/10.1007/s43621-024-00736-y.

Khodjaev, Shovkat, Ihtiyor Bobojonov, Lena Kuhn, and Thomas Glauben. 2025. "Optimizing Machine Learning Models for Wheat Yield Estimation Using a Comprehensive UAV Dataset." *Modeling Earth Systems and Environment* 11 (1): 15. https://doi.org/10.1007/s40808-024-02188-9.

Kinachtchouk, Nickolas, and David Canes. 2025. "Artificial Intelligence (AI) and Men's Health Clinic Efficiency and Clinic Billing." *Current Urology Reports* 26 (1): 23. https://doi.org/10.1007/s11934-024-01252-2.

Kong, Jiashu, Yitong Dong, Zhechen Zhang, Pow-Seng Yap, and Yuekuan Zhou. 2024. "Advances in Smart Cities with System Integration and Energy Digitalization Technologies: A State-of-the-Art Review." *Sustainable Energy Technologies and Assessments* 72 (December): 104012. https://doi.org/10.1016/j.seta.2024.104012.

Kurata, Lehlohonolo, Musa Adekunle Ayanwale, Rethabile Rosemary Molefi, and Tajudeen Sanni. 2025. "Teaching Religious Studies with Artificial Intelligence: A Qualitative Analysis of Lesotho Secondary Schools Teachers' Perceptions." *International Journal of Educational Research Open* 8 (June): 100417. https://doi.org/10.1016/j.ijedro.2024.100417.

Lee, Sangryeong, Soo-Kyoung Lee, and Jin-Woo Park. 2024. "The Effect of Service Quality and Sustainability Practices on Brand Equity: The Case of Korean Air Passengers." *Sustainability* 16 (11): 4606. https://doi.org/10.3390/su16114606.

Li, Chuang, Qin Zhao, and Liping Wang. 2024. "The Impact Mechanism and Effect Evaluation of Digital Economy Development on Regional Carbon Emission Reduction: Evidence from Provincial Panel Data in China." *International Journal of Environmental Research* 18 (6): 99. https://doi.org/10.1007/s41742-024-00651-0.

Lorenc, Augustyn, and Małgorzata Kuźnar. 2024. "Minimization of Material Waste through Maintenance Interval Optimization in Transport Systems." *Applied Sciences* 14 (23): 11318. https://doi.org/10.3390/app142311318.

Maalouf, Emmanuel. 2024. "Achieving Corporate Environmental Responsibility through Emerging Sustainability Laws." *Asia Pacific Journal of Environmental Law* 27 (1): 64–99. https://doi.org/10.4337/apjel.2024.01.03.

Magd, Hesham, and Ilya Bystrov. 2021. "Commitment of Higher Educational Institutions to Sustainability Agenda: Comparative Study of GCC and ASEAN." *Global Business and Management Research: An International Journal* 13 (3).

Medaglia, Rony, Boriana Rukanova, and Ziyan Zhang. 2024. "Digital Government and the Circular Economy Transition: An Analytical Framework and a Research Agenda." *Government Information Quarterly* 41 (1): 101904. https://doi.org/10.1016/j.giq.2023.101904.

Mehta, Dhyani, and Mohd Asif Shah. 2024. "BRICS Carbon Emissions: Asymmetric Impact of Energy Mix, Financial Development, and Digitalization." *Sustainable Environment* 10 (1): 2418162. https://doi.org/10.1080/27658511.2024.2418162.

Merlo, Tereza Raquel. 2024. "Emerging Role of Artificial Intelligence (AI) in Aviation: Using Predictive Maintenance for Operational Efficiency." In *Advances in Mechatronics and Mechanical Engineering*, edited by Ayse Asli Yilmaz, 28–46. IGI Global. https://doi.org/10.4018/979-8-3693-0732-8.ch002.

Murray, Alan, Keith Skene, and Kathryn Haynes. 2017. "The Circular Economy: An Interdisciplinary Exploration of the Concept and Application in a Global Context." *Journal of Business Ethics* 140 (3): 369–80. https://doi.org/10.1007/s10551-015-2693-2.

Nagiah, Ganesh Rao, and Norazah Mohd Suki. 2024. "Linking Environmental Sustainability, Social Sustainability, Corporate Reputation and the Business Performance of Energy Companies: Insights from an Emerging Market." *International Journal of Energy Sector Management* 18 (6): 1905–22. https://doi.org/10.1108/IJESM-06-2023-0003.

Nassar, Ahmed K. 2024. "Strategic Energy Transition in the Gulf Cooperation Council: Balancing Economic, Social, Political, and Environmental Dynamics for Sustainable Development." *International Journal of Green Energy*, December, 1–17. https://doi.org/10.1080/15435075.2024.2435434.

Obrecht, Matevž, Tina Cvahte Ojsteršek, and Lazar Pavić. 2024. "Sustainable Synergy: Unveiling the Symbiotic Relationship between Personal Lifestyles and Corporate Sustainable Business Practices." *Energy Reports* 12 (December): 6093–6101. https://doi.org/10.1016/j.egyr.2024.11.090.

Paridhi, and Ashima Arora. 2023. "Sustainability Reporting: Current State and Challenges." *Business Strategy & Development* 6 (3): 362–81. https://doi.org/10.1002/bsd2.244.

Pérez-Mesa, Juan Carlos, Laura Piedra-Muñoz, Ma Carmen García-Barranco, and Cynthia Giagnocavo. 2019. "Response of Fresh Food Suppliers to Sustainable Supply Chain Management of Large European Retailers." *Sustainability* 11 (14): 3885. https://doi.org/10.3390/su11143885.

Piracci, Giovanna, Fabio Boncinelli, and Leonardo Casini. 2022. "Wine Consumers' Demand for Social Sustainability Labeling: Evidence for the Fair Labor Claim." *Applied Economic Perspectives and Policy* 44 (4): 1742–61. https://doi.org/10.1002/aepp.13260.

Radicic, Dragana, and Saša Petković. 2023. "Impact of Digitalization on Technological Innovations in Small and Medium-Sized Enterprises (SMEs)." *Technological Forecasting and Social Change* 191 (June): 122474. https://doi.org/10.1016/j.techfore.2023.122474.

Shen, Yongtao, and Boonsub Panichakarn. 2023. "Sustainability of Agricultural Trade Supply Chains Status, Opportunities, and Future Directions." *Uncertain Supply Chain Management* 11 (4): 1769–82. https://doi.org/10.5267/j.uscm.2023.6.014.

Singhania, Monica, and Neha Saini. 2023. "Institutional Framework of ESG Disclosures: Comparative Analysis of Developed and Developing Countries." *Journal of Sustainable Finance & Investment* 13 (1): 516–59. https://doi.org/10.1080/20430795.2021.1964810.

Tagscherer, Florian, and Claus-Christian Carbon. 2023. "Leadership for Successful Digitalization: A Literature Review on Companies' Internal and External Aspects of Digitalization." *Sustainable Technology and Entrepreneurship* 2 (2): 100039. https://doi.org/10.1016/j.stae.2023.100039.

Tasci, Asli D.A. 2017. "Consumer Demand for Sustainability Benchmarks in Tourism and Hospitality." *Tourism Review* 72 (4): 375–91. https://doi.org/10.1108/TR-05-2017-0087.

Teow, Hui Hui, Pervaiz K. Ahmed, Mahendhiran Sanggaran Nair, and Santha Vaithilingam. 2024. "Green Education Divide: A Comparative Case Study Analysis of Countries of the Association of Southeast Asian Nations." *The Lancet Planetary Health* 8 (April): S20. https://doi.org/10.1016/S2542-5196(24)00085-8.

Thai, Huy-Tan, and Kim-Hung Le. 2025. "MobileH-Transformer: Enabling Real-Time Leaf Disease Detection Using Hybrid Deep Learning Approach for Smart Agriculture." *Crop Protection* 189 (March): 107002. https://doi.org/10.1016/j.cropro.2024.107002.

Ting, S. L., Y. K. Tse, G. T. S. Ho, S. H. Chung, and G. Pang. 2014. "Mining Logistics Data to Assure the Quality in a Sustainable Food Supply Chain: A Case in the Red Wine Industry." *International Journal of Production Economics* 152 (June): 200–209. https://doi.org/10.1016/j.ijpe.2013.12.010.

Umair, Sonia, Umair Waqas, Beata Mrugalska, and Ibrahim Rashid Al Shamsi. 2024. *Human Perspectives of Industry 4.0 Organizations: Reviewing Sustainable Performance*. 1st ed. CRC Press. https://doi.org/10.1201/9781032616810.

Umar, Tariq, and Nnedinma Umeokafor. 2022. "Exploring the GCC Progress towards United Nations Sustainable Development Goals." *International Journal of Social Ecology and Sustainable Development* 13 (1): 1–32. https://doi.org/10.4018/IJSESD.2022010105.

Wang, Ruoyun, Corey Kewei Xu, and Xun Wu. 2024. "Open Government Data (OGD) as a Catalyst for Smart City Development: Empirical Evidence from Chinese Cities." *Government Information Quarterly* 41 (4): 101983. https://doi.org/10.1016/j.giq.2024.101983.

Waqas, U., S. Umair, and B. Mrugalska, Ibrahim Rashid Al Shamsi, and Ilya Bystrov. 2024. "Mediating Role of Green Talent Management between Green Strategic Orientation and Sustainable Supply Chain Performance among SMEs of Oman." *Annals of Operations Research* 340 (2–3), 1063–89. https://doi.org/10.1007/s10479-024-06159-2.

Ward, James D., Paul C. Sutton, Adrian D. Werner, Robert Costanza, Steve H. Mohr, and Craig T. Simmons. 2016. "Is Decoupling GDP Growth from Environmental Impact Possible?" Edited by Daniel E. Naya. *PLOS One* 11 (10): e0164733. https://doi.org/10.1371/journal.pone.0164733.

Whig, Pawan, Balaram Yadav Kasula, Nikhitha Yathiraju, Anupriya Jain, and Seema Sharma. 2024. "Transforming Aviation: The Role of Artificial Intelligence in Air Traffic Management." In *Advances in Mechatronics and Mechanical Engineering*, edited by Saifullah Khalid and Neha Nazneen Siddiqui, 60–75. IGI Global. https://doi.org/10.4018/979-8-3693-1954-3.ch004.

Xu, Jiaqi, Xiaowan Xu, and Gujiang Jia. 2024. "Resource Efficiency, Green Development, and Social Security: Evidence from ASEAN+6 Economies." *Resources Policy* 90 (March): 104744. https://doi.org/10.1016/j.resourpol.2024.104744.

Zhang, Hanzhi, Jingfeng Zhang, and Chih-Hung Pai. 2024. "Promoting Carbon Neutrality and Green Growth through Cultural Industry Financing." *Humanities and Social Sciences Communications* 11 (1): 71. https://doi.org/10.1057/s41599-023-02554-5.

Zhang, Qixuan, Yuxin Yang, Xue Li, and Pingping Wang. 2024. "Digitalization and Agricultural Green Total Factor Productivity: Evidence from China." *Agriculture* 14 (10): 1805. https://doi.org/10.3390/agriculture14101805.

Zhao, Yingmei, and Wenping Wang. 2025. "The Multiple Empowerment Effects of Digital Transformation on Carbon Emissions in Manufacturing Industry from the Prospective of Factor Allocation: Theoretical Analysis and Empirical Evidence." *Environmental Impact Assessment Review* 110 (January): 107698. https://doi.org/10.1016/j.eiar.2024.107698.

Zhao, Zhihua, and Zhenjiang Xing. 2024. "Corporate Management, Green Finance, and Sustainability." *Humanities and Social Sciences Communications* 11 (1): 6. https://doi.org/10.1057/s41599-023-02490-4.

Zhou, Yuekuan, and Jiangyang Liu. 2024. "Advances in Emerging Digital Technologies for Energy Efficiency and Energy Integration in Smart Cities." *Energy and Buildings* 315 (July): 114289. https://doi.org/10.1016/j.enbuild.2024.114289.

Zou, Yonghua, Ni Zhong, Ziwei Chen, and Wanxia Zhao. 2024. "Bridging Digitalization and Sustainability in Universities: A Chinese Green University Initiative in the Digital Era." *Journal of Cleaner Production* 469 (September): 143181. https://doi.org/10.1016/j.jclepro.2024.143181.

Zou, Yuxuan, Nengmin Wang, Rong Dong, and Zhengwen He. 2025. "Shareable or Exclusive: The Decision Regarding Green Technological Innovation." *Journal of Industrial and Management Optimization* 21 (1): 42–78. https://doi.org/10.3934/jimo.2024075.

8 Navigating the Metaverse
Transforming Healthcare Supply Chains

Zeeshan Asim

8.1 INTRODUCTION

The metaverse, a concept first articulated by visionary author Neal Stephenson, denotes a sophisticated virtual realm that converges with reality in a manner that is increasingly imperceptible to the human experience (Stephenson, 1994). In recent years, the concept of the metaverse has gained significant attention, with many industries exploring its potential applications (Dwivedi et al., 2022). The healthcare sector, characterized by intricate networks of providers, manufacturers, and distributors, presents a fertile ground for the metaverse to unleash its transformative potential, thereby recalibrating the very fabric of its logistics and procurement systems (Koohang et al., 2023). The healthcare supply chain is a complex network of stakeholders, including manufacturers, distributors, hospitals, and patients, that work together to deliver medical products and services (Choi et al., 2021). However, this supply chain is often plagued by inefficiencies, including delays, stockouts, and counterfeiting, which can have serious consequences for patient health and safety (KPMG, 2023).

The metaverse offers a potential solution to these problems by enabling the creation of a virtual replica of the healthcare supply chain. A mirror-image simulation, or "digital twin," (DT) of the supply chain can be harnessed to orchestrate a synergistic fusion of predictive analytics, machine learning (ML), and real-time data, thereby enabling proactive decision-making, agile inventory management, and seamless logistics synchronization (Zhen-Song & Jie-Qun, 2024). Furthermore, the metaverse can serve as a catalyst for the integration of cutting-edge technologies, including distributed ledger systems, cognitive computing, and ubiquitous networking, thereby fostering a paradigm shift in supply chain visibility, resilience, and performance (IBM, 2023). For example, blockchain can be used to track and verify the authenticity of medical products, while artificial intelligence (AI) can be used to predict and prevent stockouts and delays (Vijaykumar et al., 2024).

Currently the concept of the metaverse has gained significant traction, precipitating a nascent interest in its potential applications (Bansal et al., 2022). Briefly, the metaverse is frequently conceptualized as a prospective iteration of the internet, wherein the boundaries between physical and virtual reality (VR) are increasingly blurred, yielding a novel paradigm for human interaction and experience (Chengoden et al., 2023). By leveraging the immersive capabilities of augmented

DOI: 10.1201/9781003611905-11

reality (AR) (Asim et al., 2025), VR, and extended reality (XR), individuals will be able to navigate the virtual realms of the metaverse with a fluidity and intuitiveness analogous to that of browsing online platforms with a graphical user interface today (Mystakidis, 2022).

Despite the potential benefits of the metaverse in healthcare supply chains, there are several challenges and problems that must be addressed (Pei Wu et al., 2024). One major challenge is the lack of standardization and interoperability between different healthcare systems and technologies (Pei Wu et al., 2024). This can make it difficult to integrate the metaverse with existing supply chain systems and technologies. Another challenge is the need for robust cybersecurity measures to protect against hacking and data breaches (Musamih et al., 2022). Additionally, there are concerns about the potential impact of the metaverse on the healthcare workforce, including the need for new skills and training (Musamih et al., 2022).

For this purpose, this article is structured as follows: the key concepts of technologies enabling metaverse-driven supply chains are presented in Section 8.2. While enhancing supply chain visibility and transparency are presented in Section 8.3, the potential benefits of the metaverse in healthcare logistics are presented in Section 8.4. The potential challenges are summarized in Section 8.5. A brief future outlook is illustrated in Section 8.6. Next, the significance of virtual worlds and VR for education is not relevant to this article and has been removed. The conclusions are in Section 8.7.

8.2 METAVERSE

The concept of the "metaverse" is born from Neal Stephenson's 1992 novel *Snow Crash*, which presents a new paradigm of immersive, digital interaction. In this VR, avatars communicate and collaborate with one another and with digital entities. What happens next (Bag et al., 2023; Bansal et al., 2022; Zhen-Song & Jie-Qun, 2024)?

The idea has seen considerable growth and improvement and now encompasses a much wider selection of inventive technologies and experience-rich formats, the very latest of which seem almost always boundless in their opportunities (Peukert et al., 2022). The "metaverse" stands for a supposed VR space in which user interactions with each other and digital objects are smooth and without boundaries space fostering a renewed paradigm of immersive, interactive experience. This is what the VR company Meta, formerly known as Facebook, has in mind for the term (Peukert et al., 2022).

The notion of a metaverse has become more apparent through its representation in science fiction. Works in this genre often depict the metaverse as a seamless, advanced digital universe, one where the physical and digital worlds are so intermingled that it's hard to tell them apart (Bag et al., 2023). These stories tend to highlight the almost utopian nature of being able to exist within a series of interconnected virtual environments (Dolgui & Ivanov, 2023). The concept of the metaverse has taken hold lately and is gathering a mighty wave of interest among tech entrepreneurs and investors, who have only recently begun to discourse about its immense possibilities (Wiederhold, 2022).

These platforms construct the incredibly immersive experience that is nearly effortless in merging the physical and virtual worlds. They generate an environment that is all but seamless between social media and e-commerce. And they do all of this while still managing to be super unprofitable (Chengoden et al., 2023). The metaverse has a rich ancestry, with its conceptual foundation being explored and iterated upon in various media forms, including literature, film, and interactive digital experiences, over a long period (Shin et al., 2024; Wiederhold, 2022). An ever-stronger fascination with the metaverse seems to be at least partly due to the highly symbiotic convergence of technological innovation—especially immersive computing and spatial visualization technology—and the burgeoning digital social ecosystem, replete with an expanding suite of online gaming platforms (Wiederhold, 2022). Still, the idea of "the metaverse" itself—certainly as a 3D virtual/augmented world type of environment, as opposed to a 2D desktop-centric one—seems poised to become a highly marketable concept (Belk et al., 2022). One reason is its intrinsic sense of being an interconnected multiverse of virtual worlds "out there." Another is its promising potential as an individual-customizable environment "in here" (Cheng et al., 2022). Yet another is its offering of a couple of free homogenous pathways through the 3D virtual environment (Mayer et al., 2023): the first pathway for intuitive exploration and interaction with a lot of different entities, including other users, and the second for exploration at a more profound level of individual experience (Cheng et al., 2022).

The individual user's unique predilections and preferences could be faithfully reflected in the customized avatars and virtual environments that they painstakingly build and Tenda reflection that renders them notwithstanding, since those environments are themselves spaces of authentic agency and self-expression, a space wherein not only the real world but also the "build world" is a virtual one (Mayer et al., 2023). This virtual center of innovation would be a powerful driver of collaboration. It would empower our users to harness their collective creativity and join in the kind of co-creative work that produces high-end, one-of-a-kind interactive experiences (Jauhiainen, 2024). Those experiences might be anything from a richer virtual presence in a meeting to something more akin to performance art. They might use our full sensory suite and push the new virtual space to its limits.

A multitude of industries and disciplines apply metaverse technology. It holds the potential to deliver unprecedented growth in exploration and discovery. Some call the metaverse potential "revolutionary" or even "transformational." But in what way is the metaverse applied, and in which industries is it likely to be applied in the future (Uddin et al., 2024)?

The metaverse serves as a catalyzer for collaborative synergy, creating an ecosystem of supply chain operations that is transparent and efficient (Mystakidis, 2022). It integrates seamlessly with DT technologies, which allows real-time performance monitoring and simulated testing of innovations. The DT is "a virtual model of a process, product, or service." These models are used for system design and optimization, usually during the testing phase. The moves are calibrated to allow "as close to perfect as possible" a one-to-one mapping between the physical entity and the virtual twin. This enables the metaverse to facilitate data-driven decision-making (Pedro et al., 2023).

8.3 OPTIMIZING HEALTHCARE SUPPLY CHAIN THROUGH EMBRACING DIGITAL TECHNOLOGIES

The healthcare industry is currently experiencing a revolutionary transformation, known informally as "Healthcare 4.0," that denotes a fundamental shift to a far more technologically advanced and interlinked healthcare ecosystem (Zhen-Song & Jie-Qun, 2024). Pioneering innovations make up strategic endeavors within the healthcare supply chain ecosystem. These innovations are pushing the shift toward an interconnected healthcare system. The technologies driving this change are distributed ledger systems, AI, DTs, and cloud-based infrastructure. These technologies make the systems we use in healthcare more agile. They also help make those systems work better with each other, which is a big part of creating a more interconnected healthcare landscape (Chen et al., 2024). The emergence of these state-of-the-art information technologies has brought about a shift in the healthcare supply chain, yielding remarkable transformative benefits. These technologies have the potential to change the very nature of the medical supply chain itself (Aslam et al., 2024). More specifically, these benefits result in substantial improvements in productivity for care teams, a clear reduction in inventory discrepancies, better billing accuracy, and an overall increase in the quality of patient-centric service. Together, these benefits produce a much more streamlined, efficient, and patient-centered healthcare model (Naureen, 2024).

The healthcare industry's speeding digital transformation is set to bring about an unprecedented seismic shift in the industry's transparency and accountability (Anurogo & Hidayat, 2024). Hospitals are well-positioned to maximize the benefits of enhanced data visualization and advanced analytics capabilities. These now allow for a higher level of informed decision-making and more optimized resource allocation and have improved patient data management (Liu et al., 2023). The digital age also presents a series of unparalleled opportunities to enrich the elevation of the patient experience. From telemedicine to portal access, the "digital" is allowing for a level of personal interaction not present before (Ageron, 2020). The digital revolution offers a chance to bring more "empathy" into the healthcare equation. Beyond just patient interaction, the digital revolution provides avenues to enrich supply chain management (SCM). More than anything else, this has been an area ripe for innovation (Suali et al., 2024). AI and big data analytics have revolutionized the way hospitals handle supply chain efficiency. We are now able to offer patient-centered care in a way not seen (Ehie & Ferreira, 2019).

At the same time, an optimized framework allows the healthcare sector to nimbly and accurately respond to unforeseen disruptions, such as sudden shortages of crucial medical supplies during public health emergencies like widespread epidemics or pandemics (Santus et al., 2021). These optimized frameworks are not just theoretical; they have real applications in today's healthcare structures. Emerging technologies such as the Internet of Things (IoT), AI, blockchain, and big data serve as the building blocks of the healthcare supply chain of the near future. Their integration generates a paradigm shift wherein the convergence of material flows and information streams creates an unprecedented level of efficiency and operational performance (Zhen-Song & Jie-Qun, 2024)

The healthcare supply chain is underperforming and at risk of replicating the pattern of lagging behind other industries when it comes to digital transformation (Marianna et al., 2024). Despite the potential of digital innovation to revolutionize this area, many long-standing challenges persist (Martin & Omar, 2021). These are so deeply ingrained that they threaten to slow down the progress we are trying to make toward the goal of optimizing resource allocation, which is a vital necessity if we're going to hit the targets of healthcare adaptability and fiscal sustainability (Tugce et al., 2019). Historically, the healthcare supply chain has concentrated on the procurement end and has not paid nearly as much attention to logistics and distribution, which are essential and elementary functions of any supply chain (Qvarfordt & Lagrosen, 2024). Still, even this basic effort in SCM is largely unrecognized and poorly performed in logistics, even though the performance of the supply chain is a significant determinant of the overall clinical effectiveness and cost value of the organization. The next area of concern is that, although many in the supply chain field recognize that these are deeply rooted problems that are hard to solve, they have offered almost no actionable ideas on how to fix them (Bhavana et al., 2018; Laith et al., 2023).

The healthcare supply chain encompasses a one-of-a-kind triad of complexities, which includes an enormous variety of medical products that have but an identical set of characteristics (Painuly et al., 2023). Elsewhere, we might navigate a relatively fixed set of policies; in the healthcare supply chain, we must pivot across a dynamic policy landscape as we respond to shifting market imperatives. Meanwhile, we contend with a web of intricate relationships among the multitude of stakeholders—hospitals, pharmaceutical firms, logistics providers, storage facilities, and others—who collectively demand a high degree of adaptability, flexibility, and coordination across our seamless operations (Ashwanim et al., 2023).

For that reason, traditional linear supply chain models often fall short of meeting the unique demands of the healthcare sector. Instead, the sector necessitates the adoption of supply chains that are responsive, agile, and architecturally capable of navigating the complex, dynamic landscape of the industry (Marianna et al., 2024). Sullivan and Staib (2018) adopt a sunny perspective on the prospect of hospital digitization, emphasizing its potential to induce innovation and operational efficiencies throughout the healthcare ecosystem and, under the direction of clinical leadership, to transform a digital disruption that is largely beyond hospital control into a manageable force that changes things for the better and changes them in a predictable way. As per Senbekov et al. (2020), the incorporation of digital technology holds the promise of substantial payoffs. These may well go far beyond just improving the immediate treatment of patients. The seminal work of Sebastian et al. (2020) brings clarity to the essential function that digital systems play in managing specimens, exchanging data without friction, and adhering to standardized operating protocols as digitization sweeps through our nation's hospitals. Marie et al. (2023) emphasize that digital technologies have generated an enormous accumulation of data, empowered decision-making, and enhanced the quality of care. This has instigated a worldwide shift toward digital hospital transformation, propelled by the many-layered advantages of digitalization. Stressing the need for an effective method of managing vendor partnerships, Thomas et al. (2020) offered the following recommendations to ensure that upstream suppliers and downstream service providers in the healthcare supply chain ecosystem are in sync and that services are delivered seamlessly: 1. Diversify upstream suppliers as a risk mitigation strategy, 2. Monitor and evaluate supplier performance continuously and seamlessly, and 3. Institute supplier development programs for key suppliers.

In Pedro et al. (2023) view, collectively, the avant-garde digital solutions, such as the IoT, AI, and blockchain technology, as well as VR/AR, represent the most efficacious strategy for dealing with the comprehensive, integrated, and decentralized problem space of healthcare SCM. This would create a synergistic and patient-centric supply chain. Kumar et al. (2022) harnessed the potentially disruptive power of AI to craft a state-of-the-art hospital e-logistics information system—one that not only affords real-time visibility into the logistics of the healthcare supply chain but also allows for the kind of actionable, high-level insights that can truly optimize the operational efficacy of that supply chain and drive sustained, year-over-year improvements in its efficiency. Dixit et al. (2019) performed groundbreaking work that significantly advanced the transparency and agility of the healthcare supply chain. Their effort was largely an outcome of the synergistic integration of cloud-based technologies and IT-enabled collaborative frameworks—supply chain enablers, as we call them.

8.4 TECHNOLOGIES THAT FACILITATE METAVERSE AT HEALTHCARE

The metaverse embodies a revolutionary digital framework integrating the capabilities of the IoT, cloud computing, AI, blockchain, VR, and AR to converge the virtual and physical worlds (Hatami et al., 2024; Zeeshan et al., 2024a). The real-time data captured through the IoT and associated sensor technologies feed into a newly built infrastructure in the cloud (Srilakshmi et al., 2024). Here, human-like AI systems process the data to generate new narratives that form the basis of adaptive, personalized experiences in the virtual realm (Mohsen et al., 2023). Secure transactions with guaranteed integrity, a necessity in the supply chain, are managed in a decentralized way, using blockchain. Instead of seeing flat computer screens, users now interact with virtual environments through a wide range of devices and immersive technologies, including head-mounted displays (Zeeshan et al., 2024b).

8.4.1 Role of IoT and Cloud Computing in Healthcare Supply Chain

The healthcare supply chain has been turned upside down by the IoT and cloud computing (Rejeb et al., 2023). This is because the IoT makes it so easy to integrate data from devices that you can capture and store real-time data across too many locations to list, and that includes hospitals, warehouses, and virtually the whole transportation network (Phillip et al., 2022; Sodhro et al., 2018). Yet all these inhuman devices are not always the answer to better supply chain performance (Reza et al., 2021). Just consider what happens when you arrive at the periphery of the network and all its edge nodes: inside hospitals, next to your medication, and in the fancy contraptions that always seem to go inside the refrigerator unit (ElRahman & Alluhaidan, 2021). Once you get to the edge, you've got to secure the data that you're sending back through the network and ensure that it remains confidential and accessible, all while also sending the processed data to some cloud server (Das et al., 2022; El Azzaoui et al., 2022).

8.4.2 Digital Twins in Healthcare Supply Chain

Visibility and transparency in the healthcare supply chain have long eluded even the best supply chain managers (Krishnaveni & Geetha, 2025). A potential solution lies in the use of DTs, which creates a virtual model of a healthcare supply chain component that closely mirrors its real-world counterpart and, when necessary, uses a "right-time, right-speed" cloud-stored data synchronization to supplement any lacking virtual model intelligence (Marianna et al., 2024; Yu et al., 2024).

DTs provide deep insights into supply chain operations, enabling the identification of risks and inefficiencies (Zhen-Song & Jie-Qun, 2024). They are employed to model vital supply chain components such as the production of medical devices, the manufacture of pharmaceuticals, warehousing, and transportation and delivery services (Saroj et al., 2023). The virtual models created using DTs closely mirror the physical world, thus permitting the human operators' well-ingestible insights into what's happening right now and what may happen next (Armeni et al., 2022).

The DTs assist managers in evaluating resilience, uncovering bottlenecks, and fine-tuning logistics, inventory control, and quality assurance to products (Yu et al., 2024). They do this by simulating barriers that are encountered when operating a supply chain (Herden et al., 2020). The updates that make the DTs "continuous" are what set them apart from the more traditional DT concept (Mertens et al., 2024). The healthcare supply chain gets better efficiency, reliability, and flexibility with the simulation capabilities of DTs, assuring higher service quality and improved patient care

8.4.3 Role of AI in Healthcare Supply Chain

Complex processes, from production to distribution and end-service delivery, make up the healthcare supply chain. AI—comprising ML, deep learning, and natural language processing (NLP)—enhances supply chain operations (Maleki Varnosfaderani & Forouzanfar, 2024). AI and its tools integrate vast amounts of data generated by IoT and cloud computing into the healthcare supply chain (Mani et al., 2022). That data and its dimensions are used to turbocharge several operations—demand forecasting, tracking, and distribution of supplies and medicine—within the supply chain (Seyedan & Mafakheri, 2020).

Demand forecasting is critically influenced by ML, which digs into historical and recent data to unearth the kinds of things that help us understand what kind of demand is coming and when (Maryam et al., 2022). And then there is logistics. ML in logistics has the same aim as in demand forecasting: to understand what is happening currently and what is likely to happen next (Abbas et al., 2020). Quality control and assurance of product authenticity are ensured by deep learning using advanced recognition techniques (Krishnaveni & Geetha, 2025). These techniques have a high rate of accuracy and can be trusted to prevent counterfeit goods from entering the marketplace (Mei et al., 2023). When applied to the supply chain, deep learning assists with data mining and analytics. It reveals hidden supply chain patterns that are critical for knowledge discovery (Azadi et al., 2023)

Medical document processing is automated by NLP, with the extraction of key information from documents rendered more efficiently (Hossain et al., 2021). When the rubber hits the road, this is what really matters. Even better, NLP is applied to decision support, with the result being more-informed decisions (Ucar et al., 2024). The integration of AI into healthcare SCM is promising to make it far more efficient, effective, and reliable and to radically improve the way it operates and the way it serves patients

8.4.4 Role of Blockchain in Healthcare Supply Chain

The metaverse healthcare supply chain can leverage blockchain technology, which offers a new way to manage and transmit data (Ali et al., 2023). Blockchain comprises units, known as blocks, that contain transactional data, linked together in a chain and ensuring and authenticating the data's integrity (Sarode et al., 2021). Given the complexity of the healthcare supply chain, high-speed tracing and secure transmissions of medicines and medical devices are essential, and this is where the promise of blockchain lies (Alessandra et al., 2024). For one thing, its structure—a distributed ledger that's immutable—creates a decentralized, tamper-proof way of storing and managing data (Papadopoulos et al., 2022). Compared to centralized systems, where everything hangs together like a house of cards, if any one card is removed or altered, the whole thing collapses, a blockchain system has no single point of failure, making it more secure and reliable (Siful & Kutub, 2024).

Managing contracts and payments in the healthcare supply chain can be optimized with blockchain (Jayendra & Jyoti, 2022). Smart contracts is a key feature of blockchain technology that ensure secure and transparent service-based payments (Manimuthu et al., 2022). These predefined codes execute actions like inventory updates and quality control (among others) when conditions are met, reducing the number of intermediaries involved too almost none (Musamih et al., 2022). Also, when it comes to medical devices, blockchain allows for real-time tracking, meaning that anything that can be done with a standard, auditable, tamper-proof, and almost instant digital agreement can also be done with a medical device (Musamih et al., 2022).

8.4.5 Role of VR/XR in Healthcare Supply Chain

The appearance of VR and AR in healthcare has started a shift in the supply chain toward interactive experiences for patients, doctors, and vendors (Tarun et al., 2023). By utilizing VR headsets and smart glasses, users can take an interactive journey through the components of the supply chain, from the preparation of medical products to the logistics of getting them to doctor and hospitals (Asim et al., 2025; Renu, 2021; Zeeshan et al., 2024a). Not only do VR and AR allow for an unprecedented level of interaction with the supply chain, but they also provide a way for users to understand and appreciate the complexities involved in its management (Akbari et al., 2022; Zhen-Song & Jie-Qun, 2024). These technologies stand to revolutionize supply chain education in the same way they have in healthcare pedagogy, through simulation of complex processes.

A more advanced concept, XR, has emerged and is the result of VR and AR technologies. XR can combine 3D virtual worlds with 3D interactive experiences in the real world, a combination that allows for a new level of immersion and interactivity (Ahmed et al., 2022; Brozovsky et al., 2024). In the healthcare SCM system of the future, XR will provide the unprecedented ability to visualize the entire healthcare supply chain in 3D and to perform interactive 3D experiences in the real world along with the digital, virtual world (Ali et al., 2023).

8.4.6 Cyber-Physical System (CPS)

In term of robust information network in case of healthcare supply chain, the CPS (Wolf, 2009) brings together networking, computing, elements from the physical world, and components of society (Lesch et al., 2023). This paradigm not only lays the groundwork for a robust metaverse healthcare supply chain but also sheds light on the CPS-based metaverse-related healthcare supply chain across three key dimensions: the cyber system, the physical system, and the social system (Senthil Kumar & Mohamed, 2022). This dimension goes beyond simple technical integration and takes a holistic systems-thinking approach. Its main aim is to unpick the traditional healthcare supply chain and to find synergies among information, material resources, and social components of the system (Senthil Kumar & Mohamed, 2022).

8.5 METAVERSE FOR HEALTHCARE SUPPLY CHAIN

A complex and interdependent set of stakeholders forms the healthcare supply chain, which exists to provide seamless medical care to the citizens. At the core of the supply chain are four major stakeholders, which can be collectively referred to as the "pivotal constituents of the healthcare value chain": 1) supply chain providers, 2) medical distribution facilities 3) health logistics operations, and 4) hospitals and the healthcare infrastructure (Zhen-Song & Jie-Qun, 2024). These four groups form the core of the supply chain, and in unison, their stakeholder components represent the main entry points for the essential products, services, and solutions that healthcare citizens require.

8.5.1 Medical Supplies Manufacturer's

The healthcare supply chain is supported by the healthcare industry's foundational entities, which consist of a diverse collection of manufacturers and suppliers; these entities serve as the linchpin of the supply chain. They produce and produce an array of medical products and services (Zhen-Song & Jie-Qun, 2024). The quintessential organizations of the healthcare sector—an array of medical and pharmaceutical producers and suppliers—lie at the epicenter of the healthcare value chain. They form the core of an industry that converges on products, services, and solutions to serve the needs of patients, healthcare professionals, and payers (Mueen et al., 2021).

In all their daily endeavors, these entities are required to maintain an unwavering loyalty to demanding quality standards and exacting regulatory stipulations. They accomplish all of this, day in and day out, to ensure that the products we receive from them are safe and effective, potent and reliable, life-sustaining, in all instances, and dangerous only when necessary ((Musamih et al., 2022; Peukert et al., 2022). One must acknowledge that healthcare organizations must prove their competence in the adaptability department—this is, after all, a "dynamic" marketplace of burgeoning demands, necessitating fast and flexible responses to emergent situations that are sure to put "pressure" on any number of components that make up their "indexed" service constellation. We are not just talking about hospital emergency departments here but all kinds of healthcare outpatient and "after-care" service venues (Zhen-Song & Jie-Qun, 2024).

8.5.2 Medical Distribution Centre

As DTs increasingly integrate with assets and operational workflows, medical distribution centers must ensure the confidentiality, integrity, and availability of their sensitive information. This focus becomes especially important when considering the external logistics operations that involve working with staff and vendors who are not directly employed by the medical distribution center. Achieving this focus requires three main elements. First, medical distribution centers need to deploy robust cybersecurity countermeasures that will protect against the "never-ending, ever-increasing, and ever-changing cyber threats" we discussed earlier. Second, these centers must standardize their communication protocols in a way that allows for nearly all possible contingencies to be covered (Kuo & Choi, 2024). Finally, medical distribution centers must ensure that the decision-making process is going to be safe and secure (for only authorized personnel), without unduly hampering the workflow of the external logistics operation (Seckin & Koksalmis, 2024).

To realize the enormous promise of the metaverse, distributors must prioritize the creation of seamless data interchange and the interoperability of different systems, platforms, and architectures. This is the first and most essential step toward a "cohesive and interconnected ecosystem" that can sustain the "free flow of information and resources" (Zhen-Song & Jie-Qun, 2024). Such an ecosystem requires standardization, that is, it needs shared data formats and protocols among the different technologies that make up the metaverse.

8.5.3 Hospital Warehouse

The part played by medical logistics facilities is vital within the healthcare value chain. These facilities act as intermediary nodes that connect the manufacturers, providers, institutions, and payers of healthcare (Marques et al., 2020). They store and manage the medical equipment, disposable products, and prescription medications needed by healthcare providers and their patients (Jæger et al., 2021). And they are increasingly called upon to ensure that all the supplies necessary for good patient care are delivered to the right place at the right time—"just in time," for instance, to be used in an operating room (Jæger et al., 2021; Marques et al., 2020).

These warehouses have a fundamental responsibility of managing their inventories with great care. That is critical to their own operations and the supply chain that serves us all (Man et al., 2024). It guarantees that a range of medical essentials are kept in a variety of environments, as appropriate to each item, and for the most part, those are controlled environments. Who's in charge of all this? We are, of course; the human factor is what makes supply chains work—or, too often, not work.

There is no cessation in the supply of medical materials that are necessary for the provision of ongoing services, for the management of emergencies, or for dealing with surprise medical crises (Zhen-Song & Jie-Qun, 2024). Therefore, the healthcare supply chain must exist and must be effective. That means the warehouses that provide the supplies must serve their function and must serve it well; if they don't, the supply chain breaks and a disaster happens on the side of the patients (Peukert et al., 2022).

8.5.4 Patient Care

The supply chain of healthcare is a singular and intricate arrangement where the end consumer is the linchpin determining the quality of service at every point along the way (Qvarfordt & Lagrosen, 2024). This calls for logistics and distribution strategies that are as tailored as the orders themselves to meet the not very well understood demands and expectations of the consumer(Kohli et al., 2021). The direct and indirect healthcare system across the globe consists of a sundry set of provider types, including not just hospitals and clinics, but also pharmacies and a host of specialists, who together form the bedrock of a system that promises and delivers integrated, if not always well-coordinated, care (Kohli et al., 2021).

The reliable and consistent supply of medical products is essential for the provision of high-quality patient care (Zhen-Song & Jie-Qun, 2024). A total of 53% of hospitals reported medical product shortages in 2021, a "system breakdown that endangers patient safety," as one healthcare attorney put it. The same report stated that 70% of hospitals rely on 10 product types that are on or near the edge of being in short supply (Caulder et al., 2015).

There are also numerous products that hospital staff use without being aware that they exist only due to their being "under the radar." Hospital staff do not know when they are using something that's in short supply; they just know when a patient is in crisis, and we can't find what's needed (Badr et al., 2024).

The healthcare supply chain encompasses many kinds of products, everything from surgical devices to everyday medical supplies—that are vital to patient safety and the overall quality of care (Eugene & Yousef, 2022). These products must be available when they are needed, they must be the right products, and they must be used in the right way if human life is to be preserved and if what patients and their families experience can reasonably be called "healthcare" (Abdulaziz et al., 2024).

8.6 BARRIER IN IMPLEMENTING IN HEALTHCARE SUPPLY CHAIN

The metaverse is about to take the world by storm with its enormous potential and far-reaching consequences. The healthcare sector needs to pay close attention to the marketplace and try to comprehend and integrate the opportunities the metaverse offers. But it's not that simple. You can't just waltz into the metaverse and start offering health services, and you certainly can't do it with the same model that brought you into the VR world of second life.

8.6.1 Data Integrity in the Metaverse

A common issue in the metaverse is deep fakes. These are threatening the validity of digital information across all platforms (Dwivedi et al., 2022). Be it AR, VR, or any other digital medium, how can one be sure of the authenticity of an asset—or an interaction with that asset—that's been digitally messed with or otherwise devalued? Digital meddlers are here to stay. So, what's the metaverse to do? Allow a few (much needed, if you're being optimistic) authenticators of digital artifacts for your next virtual outfit or dance with Shakira (Wang et al., 2022)?

Despite the difficulties posed by deep fakes, it is essential to ensure the authenticity of data in the metaverse (Zaman et al., 2022). Why? Because accurate data is vitally important for patient care and diagnosis (Furlanetto et al., 2020). When accurate data is absent, it hinders a medical specialist's ability to tell the real from the fake, and this surely compromises the "efficacy of [the] treatment" (Yan & Chen, 2019). Moreover, healthcare databases are deeply affected by the "deceiving attacks" attaching to deep fakes. How do these attacks hit the databases? By compromising the "veracity of sensitive patient information." And what does this do? It makes the healthcare system far less trustworthy. It also puts the "accuracy of diagnoses and treatment plans" in jeopardy. The real data presenting challenge in the "metaverse" makes it the next big biennial to researcher avenue (Kostick-Quenet & Rahimzadeh, 2023).

8.6.2 The Imperative of Data Security in the Metaverse-Based Healthcare Supply Chain

The metaverse-based healthcare supply chain delivers a vast array of patient data and critical medical intelligence (Wang et al., 2023). This makes it even more vital to ensure that the confidential information that they share remains safe and sound. Find out how the healthcare supply chain can protect itself against data breaches. And learn what kinds of threats it faces from a cadre of cybercriminals looking to pillage and profit (Wang et al., 2023). In term of, awareness it is necessary to scale the damage done to the healthcare reputation from an incidents that can't be quantified, nor can the breach of trust we suffer with patients when we realize that we're not even keeping their most sensitive information secure (Reddy et al., 2023; Shree et al., 2022).

8.6.3 Legal Encounters in the Metaverse

A complex interplay of laws and regulations shapes the fledgling metaverse-based healthcare supply chain (Finch et al., 2023). They have far-reaching ramifications, most notably in several key areas where their effects are pronounced and multifaceted. Regulations concerning the protection of data provide a sturdy foundation for lawful processing and secure management of patient information. Patient med data is some of the most sensitive and valuable data today (Deven & Kenneth,

2021). It is imperative to establish a multi-layered security posture that incorporates state-of-the-art encryption and access control methodologies to ensure that it is handled and transported across the supply chain with the appropriate level of confidentiality and integrity (Karall, 2021). And what happens if/when something goes wrong? Well, these regulatory frameworks require that patients be notified, and quickly. And it must be done in language and terms that are easily understood. Most importantly, if the information is to remain secure, there is no room for secret handshakes in accomplishing this task: The law is clear that patients must be given a "plainly stated" account of the incident at issue.

8.6.4 Technological Hurdle among Supply Chain Entities

The existing technological infrastructure is outdated and a big obstacle to forming and integrating a healthcare supply chain that is part of the metaverse, pointing to the need for very innovative solutions to get over this technological hurdle (Majumder & Dey, 2024). A lot of people involved in the healthcare supply chain are still using old information systems and equipment that can't keep up with the technological pace of today or can't operate at a level that is anywhere close to satisfactory. And these ancient systems are a big part of the problem (Ikram & Habib, 2022).

A situation gives rise to a multitude of problems and includes several facets. Among these are the fragility of hardware systems, which are forever in danger of being made obsolete or of falling victim to the shortfalls of profound software inadequacies (Zhen-Song & Jie-Qun, 2024). These issues cause a large number of operational inefficiencies all over the healthcare supply chain, which lag the movement of materials into and throughout the many logistical processes in healthcare (Dasaklis & Tsoulfas, 2024). In addition, the outdated technological infrastructure leads to poor data integrity and material management, which worsens the operational inefficiencies and the overall ineffectiveness of the supply chain (Dasaklis & Tsoulfas, 2024).

8.7 CONCLUSION

The complex challenges that the healthcare supply chain faces go well beyond operational efficiency, and they result from—and are impacted by—a range of complex technological, organizational, and environmental factors. The metaverse-based healthcare supply chain is a potentially transformative solution, with the promise of enhanced efficiency, better traceability, and improved decision-making to resolve those challenges. This paper probes its theoretical underpinnings, using the CPSS framework, and identifies the key technologies—IoT, blockchain, AI, and VR/AR—that might collectively improve visibility, security, and automation in the supply chain. The metaverse applications in healthcare SCM face technological limitations, regulatory hurdles, and integration challenges that stem from big data and the IoT. Overcoming these will be essential for the healthcare worry supply chain to successfully function as a metaverse application. It could be one of the first instances when a full-scale framework—so needed in the virtually reality development era—is adaptively applied, in practice, across multiple diverse healthcare ecosystems. "Bridging" theory with practice means, first, delivering the metaverse supply dash with real-world data.

REFERENCES

Abbas, K., Afaq, M., Ahmed, K. T., & Song, W.-C. (2020). A blockchain and machine learning-based drug supply chain management and recommendation system for smart pharmaceutical industry. *Electronics*, *9*(5), 852. https://doi.org/10.3390/electronics9050852

Abdulaziz, A., Khalifa, N. A.-K., & Abdel, M. H. (2024). AI-powered innovation in digital transformation: Key pillars and industry impact. *Sustainability*, *16*(5), 1790. https://doi.org/10.3390/su16051790

Alnagrat, A., Ismail, R. C., & Syed Idrus, S. Z. (2022). A review of extended reality (XR) technologies in the future of human education: Current trend and future opportunity. *Journal of Human Reproductive Sciences*, *1*, 81–96. https://doi.org/10.11113/humentech.v1n2.27

Akbari, M., Ha, N., & Kok, S. (2022). A systematic review of AR/VR in operations and supply chain management: Maturity, current trends and future directions. *Journal of Global Operations and Strategic Sourcing*, *15*(4), 534–565. https://doi.org/10.1108/JGOSS-09-2021-0078

Alessandra, R., Sabrina, S., Jesus, F. C. M., & Alberto, C.-P. (2024). IoT-driven blockchain to manage the healthcare supply chain and protect medical records. *Future Generation Computer Systems*, *161*, 415–431. https://doi.org/10.1016/j.future.2024.07.039

Ali, S., Abdullah, Armand, T. P. T., Athar, A., Hussain, A., Ali, M., Yaseen, M., Joo, M.-I., & Kim, H.-C. (2023). Metaverse in healthcare integrated with explainable AI and blockchain: Enabling immersiveness, ensuring trust, and providing patient data security. *Sensors*, *23*(2), 565. https://doi.org/10.3390/s23020565

Anurogo, D., & Hidayat, N. A. (2024). *Digital Health Made Easy*. Nas Media Pustaka. https://books.google.com.om/books?id=xVAgEQAAQBAJ

Armeni, P., Polat, I., De Rossi, L. M., Diaferia, L., Meregalli, S., & Gatti, A. (2022). Digital twins in healthcare: Is it the beginning of a new era of evidence-based medicine? A critical review. *Journal of Personalized Medicine*, *12*(8), 1255. https://doi.org/10.3390/jpm12081255

Ashwanim, K., Venkatesh, M., Vranda, J., Himanshu, G., & Venkatesh, V. G. (2023). Managing healthcare supply chain through artificial intelligence (AI): A study of critical success factors. *Computers & Industrial Engineering*, *175*. https://doi.org/10.1016/j.cie.2022.108815

Asim, Z., Sorooshian, S., Al Shamsi, I. R., Muniyanayaka, D., & Al Azzani, A. (2025). *Supply Chain 4.0 a Source of Sustainable Initiative Across Food Supply Chain: Trends and Barriers*. CRC Press. https://doi.org/10.1201/9781032616810

Aslam, M. M., Zakari, R. Y., Tufail, A., Ali, S., Kalinaki, K., & Shafik, W. (2024). *Introduction to Industry's Fourth Revolution and Its Impacts on Healthcare* (Vol. 1). De Gruyter. https://doi.org/10.1515/9783111327853-002

Naureen, A., Vamshi, K., Krishna, K. C., & Hussain, M. M. (2024). Healthcare 5.0 fundamentals. In V. P. A. Hassan, P. Bhattacharya, P. Dutta, & R. Damaševičius (Eds.), *Federated learning and AI for Healthcare 5.0* (pp. 1–25). IGI Global Scientific Publishing. https://doi.org/https://doi.org/10.4018/979-8-3693-1082-3.ch001

Azadi, M., Yousefi, S., Saen, R. F., Shabanpour, H., & Jabeen, F. (2023). Forecasting sustainability of healthcare supply chains using deep learning and network data envelopment analysis. *Journal of Business Research*, *154*, 113357. https://doi.org/10.1016/j.jbusres.2022.113357

Ageron, B., Bentahar, O., & Gunasekaran, A. (2020). Digital supply chain: Challenges and future directions. *Supply Chain Forum: An International Journal*, *21*(3), 133–138. https://doi.org/10.1080/16258312.2020.1816361

Badr, Y., Abdul, K. L., & Shamayleh, A. (2024). The use of big data in personalized healthcare to reduce inventory waste and optimize patient treatment. *Journal of Personalized Medicine*, *14*(3), 383. https://doi.org/10.3390/jpm14040383

Bag, S., Rahman, M. S., Srivastava, G., & Shrivastav, S. K. (2023). Unveiling metaverse potential in supply chain management and overcoming implementation challenges: An empirical study. *Benchmarking: An International Journal*. https://doi.org/10.1108/BIJ-05-2023-0314

Bansal, G., Rajgopal, K., Chamola, V., Xiong, Z., & Niyato, D. (2022). Healthcare in metaverse: A survey on current metaverse applications in healthcare. *IEEE Access*, *10*, 119914–119946. https://doi.org/10.1109/ACCESS.2022.3219845

Belk, R., Humayun, M., & Brouard, M. (2022). Money, possessions, and ownership in the metaverse: NFTs, cryptocurrencies, Web3 and wild markets. *Journal of Business Research*, *153*, 198–205. https://doi.org/10.1016/j.jbusres.2022.08.031

Bhavana, M., Meena, M. L., Gupta, S., & Dangayach, G. S. (2018). Healthcare supply chain management: Literature review and some issues. *Journal of Advances in Management Research*, *15*(1). https://doi.org/10.1108/JAMR-09-2017-0090

Brozovsky, J., Labonnote, N., & Vigren, O. (2024). Digital technologies in architecture, engineering, and construction. *Automation in Construction*, *158*, 105212. https://doi.org/10.1016/j.autcon.2023.105212

Caulder, C. R., Mehta, B., Bookstaver, P. B., Sims, L. D., & Stevenson, B. (2015). Impact of drug shortages on health system pharmacies in the southeastern United States. *Hospital pharmacy*, *50*(4), 279–286. https://doi.org/https://doi.org/10.1310/hpj5004-279

Chen, Z.-S., Liang, C.-Z., Xu, Y.-Q., Pedrycz, W., & Skibniewski, M. (2024). Dynamic collective opinion generation framework for digital transformation barrier analysis in the construction industry. *Information Fusion*, *103*, 102096. https://doi.org/10.1016/j.inffus.2023.102096

Cheng, R., Wu, N., Chen, S., & Han, B. (2022). Will metaverse be nextg internet? Vision, hype, and reality. *IEEE Network*, *36*(5), 197–204. https://doi.org/10.1109/MNET.117.2200055

Chengoden, R., Victor, N., Huynh-The, T., Yenduri, G., Jhaveri, R. H., Alazab, M., & Gadekallu, T. R. (2023). Metaverse for healthcare: A survey on potential applications, challenges and future directions. *IEEE Access*, *11*, 12765–12795. https://doi.org/10.1109/ACCESS.2023.3241628

Choi, T. Y., Li, J. J., Rogers, D. S., Schoenherr, T., & Wagner, S. M. (2021). *The Oxford Handbook of Supply Chain Management*. Oxford University Press. https://books.google.com.om/books?id=t4HgEAAAQBAJ

Das, J., Ghosh, S., Mukherjee, A., Ghosh, S. K., & Buyya, R. (2022). RESCUE: Enabling green healthcare services using integrated IoT-edge-fog-cloud computing environments. *Software: Practice and Experience*, *52*(7), 1615–1642. https://doi.org/10.1002/spe.3078

Dasaklis, T. K., & Tsoulfas, G. T. (2024). The future of healthcare supply chains: Integrating Industry 4.0 technologies for improved resilience and sustainability. In Jawab, F. (Ed.), *Hospital Supply Chain. Integrated Science* (Vol. 27, pp. 533–551). Springer Nature. https://doi.org/10.1007/978-3-031-70292-1_25

Deven, M., & Kenneth, D. M. (2021). Privacy protections to encourage use of health-relevant digital data in a learning health system. *npj Digital Medicine*, *4*(2). https://doi.org/10.1038/s41746-020-00362-8

Dixit, A., Routroy, S., & Dubey, S. K. (2019). A systematic literature review of healthcare supply chain and implications of future research. *International Journal of Pharmaceutical and Healthcare Marketing*, *13*(4), 405–435. https://doi.org/10.1108/IJPHM-05-2018-0028

Dolgui, A., & Ivanov, D. (2023). Metaverse supply chain and operations management. *International Journal of Production Research*, *61*(23), 8179–8191. https://doi.org/10.1080/00207543.2023.2240900

Dwivedi, Y. K., Hughes, L., Baabdullah, A. M., Ribeiro-Navarrete, S., Giannakis, M., Al-Debei, M. M., & Wamba, S. F. (2022). Metaverse beyond the hype: Multidisciplinary perspectives on emerging challenges, opportunities, and agenda for research, practice and policy. *International Journal of Information Management*, *66*. https://doi.org/10.1016/j.ijinfomgt.2022.102542

Ehie, I., & Ferreira, L. M. D. (2019). Conceptual development of supply chain digitalization framework. *IFAC-PapersOnLine*, *52*(13), 2338–2342. https://doi.org/10.1016/j.ifacol.2019.11.555

El Azzaoui, A., Chen, H., Kim, S. H., Pan, Y., & Park, J. H. (2022). Blockchain-based distributed information hiding framework for data privacy preserving in medical supply chain systems. *Sensors*, *22*, 1371. https://doi.org/10.3390/s22041371

ElRahman, S. A., & Alluhaidan, A. S. (2021). Blockchain technology and IoT-edge framework for sharing healthcare services. *Soft Computing*, *25*, 13753–13777. https://doi.org/10.1007/s00500-021-06041-4

Eugene, S., & Yousef, A. (2022). Supply chain management and health services research: Aligning strange bedfellows. *Health Service Research*, *57*(2), 223–226. https://doi.org/10.1111/1475-6773.13939

Finch, H., Affia, A.-A., Jung, W., Potter, L., & Palmer, X.-L. (2023). Commentary on healthcare and disruptive innovation. *International Conference on Cyber Warfare and Security*, *18*(1), 77–84. https://doi.org/10.34190/iccws.18.1.999

Furlanetto, T. M., Luciano, E. M., Magnagnagno, O. A., & Lübeck, R. M. (2020). Information security in healthcare supply chains: An analysis of critical information protection practices. *Gestão & Produção*, *27*(4). DOI:10.1590/0104-530x5376-20

Hatami, M., Qu, Q., Chen, Y., Kholidy, H., Blasch, E., & Ardiles-Cruz, E. (2024). A survey of the real-time metaverse: Challenges and opportunities. *Future Internet*, *16*(10), 379. https://doi.org/10.3390/fi16100379

Herden, T. T., Nitsche, B., & Gerlach, B. (2020). Overcoming barriers in supply chain analytics—Investigating measures in LSCM organizations. *Logistics*, *4*(1), 5. https://doi.org/10.3390/logistics4010005

Hossain, E., Rana, R., Higgins, N., Soar, J., Barua, P. D., Pisani, A., & Turner, K. (2021). Natural language processing in electronic health records in relation to healthcare decision-making: A systematic review. *Computers in Biology and Medicine*, *155*(1), 107076. https://doi.org/10.1016/j.compbiomed.2023.106649

IBM. (2023). *IBM Blockchain Transparent Supply*. https://www.ibm.com/downloads/documents/us-en/10a99803aaafd73e

Ikram, H., & Habib, M. M. (2022). The future of supply chain management through technological advancements. *International Supply Chain Technology Journal*, *8*(6). https://doi.org/10.20545/isctj.v08.i06.01

Jæger, B., Menebo, M. M., & Upadhyay, A. (2021). Identification of environmental supply chain bottlenecks: A case study of the Ethiopian healthcare supply chain. *Management of Environmental Quality: An International Journal*, *32*(6), 1233–1254. https://doi.org/10.1108/MEQ-12-2019-0277

Jauhiainen, J. S. (2024). The metaverse: Innovations and generative AI. *International Journal of Innovation Studies*, *8*(3), 262–272. https://doi.org/10.1016/j.ijis.2024.04.004

Jayendra, S. J., & Jyoti, D. (2022). A review study of the blockchain-based health-care supply chain. *Social Sciences & Humanities Open*, *6*(1), 100328. https://doi.org/10.1007/978-3-031-18082-8_17

Karall, M. T. (2021). *Legal Aspects of Data Protection Regarding Health and Patient Data in the European Context*. Springer International Publishing. https://doi.org/10.1007/978-3-030-69262-9_31

Kohli, R., Garg, A., Phutela, S., Kumar, Y., & Jain, S. (2021). An improvised model for securing cloud-based e-healthcare systems. In G. Marques, A. K. Bhoi, V. H. C. D. Albuquerque, & H. K.S. (Eds.), *IoT in Healthcare and Ambient Assisted Living* (Vol. 933, pp. 293–310). Springer Singapore. https://doi.org/10.1007/978-981-15-9897-5_14

Koohang, A., Nord, J. H., Ooi, K. B., Tan, G. W. H., Al-Emran, M., Aw, E. C. X., & Wong, L. W. (2023). Shaping the metaverse into reality: A holistic multidisciplinary understanding of opportunities, challenges, and avenues for future investigation. *Journal of Computer Information Systems*, *63*(3), 735–765. https://doi.org/10.1080/08874417.2023.2165197

Kostick-Quenet, K., & Rahimzadeh, V. (2023). Ethical hazards of health data governance in the metaverse. *Nature Machine Intelligence*, *5*(5), 480–482. https://doi.org/10.1038/s42256-023-00658-w

KPMG. (2023). *The Future of Supply Chain* https://assets.kpmg.com/content/dam/kpmg/xx/pdf/2023/09/kpmg-future-of-supply-chain-report.pdf

Krishnaveni, S., & Geetha, M. (2025). *AI, IoT, and Blockchain in Healthcare: Bridging Technology and Patient Wellbeing*. IGI Global Scientific Publishing InfoSci. https://doi.org/10.4018/979-8-3373-0240-9.ch008

Kumar, A., Mani, V., Jain, V., Gupta, H., & Venkatesh, V. (2022). Managing healthcare supply chain through artificial intelligence (AI): A study of critical success factors. *Computers & Industrial Engineering*, *175*. https://doi.org/10.1016/j.cie.2022.108815

Kuo, H.-T., & Choi, T.-M. (2024). Metaverse in transportation and logistics operations: An AI-supported digital technological framework. *Transportation Research Part E: Logistics and Transportation Review*, *185*. https://doi.org/10.1016/j.tre.2024.103496

Laith, A., Essam, S. H., Raed, A. Z., Cuong, L. T., Samir, K., & Amir, H. G. (2023). Revolutionizing sustainable supply chain management: A review of metaheuristics. *Engineering Applications of Artificial Intelligence*, *126*. https://doi.org/10.1016/j.engappai.2023.106839

Lesch, V., Züfle, M., Bauer, A., Iffländer, L., Krupitzer, C., & Kounev, S. (2023). A literature review of IoT and CPS—What they are, and what they are not. *Journal of Systems and Software*, *200*(3), 111631. https://doi.org/10.1016/j.jss.2023.111631

Liu, P., Wang, Z., Liu, N., & Peres, M. A. (2023). A scoping review of the clinical application of machine learning in data-driven population segmentation analysis. *Journal of the American Medical Informatics Association*,*30*(9), 1573–1582. https://doi.org/10.1093/jamia/ocad111

Majumder, S., & Dey, N. (2024). Relationship between metaverse countries and Industry 5.0. In *Metaverse for Industry 5.0*. Springer. https://doi.org/10.1007/978-981-97-2455-0_7

Maleki Varnosfaderani, S., & Forouzanfar, M. (2024). The role of AI in hospitals and clinics: Transforming healthcare in the 21st century. *Bioengineering*, *11*(4), 337. Https://doi.org/10.3390/bioengineering11040337

Man, L. C., Lin, Y., Pang, G., Sanderson, J., & Duan, K. (2024). Digitalization to achieve greener healthcare supply chain. *Journal of Cleaner Production*, *463*. https://doi.org/10.1016/j.jclepro.2024.142802

Mani, D. K., Sheeba, J.I., & Pradeep Devaneyan, S. (2022). Integrated Internet of Things with cloud developed for data integrity problems on supply chain management. *Measurement: Sensors*, *24*. https://doi.org/10.1016/j.measen.2022.100445

Manimuthu, A.,.., Venkatesh, V. G., Sobhan, A., Yangyan, S., & Raja Sreedharan, V. (2022). Application of blockchain and smart contracts in autonomous vehicle supply chains: An experimental design. *Transportation Research Part E: Logistics and Transportation Review, 165*(102864). https://doi.org/10.1016/j.tre.2022.102864

Marianna, M., Guido, N., Anna, P., & Fabrizia, S. (2024). Digital transformation in healthcare: Assessing the role of digital technologies for managerial support processes. *Technological Forecasting and Social Change, 209*. https://doi.org/10.1016/j.techfore.2024.123781

Marie, W., Livia, M. S., & Janna, H. (2023). Experience of health care professionals using digital tools in the hospital: Qualitative systematic review. *JMIR Human Factors, 10*(4). https://doi.org/10.2196/50357.

Marques, L., Martins, M., & Araújo, C. (2020). The healthcare supply network: Current state of the literature and research opportunities. *Production Planning & Control, 31*(7), 590–609. https://doi.org/10.1080/09537287.2019.1663451

Martin, B., & Omar, B. (2021). Digitalization of the healthcare supply chain: A roadmap to generate benefits and effectively support healthcare delivery. *Technological Forecasting and Social Change, 167*, 120717. https://doi.org/10.1016/j.techfore.2021.120717

Maryam, Z., Majid, R., Vahid, K., & Amirhossein, S. (2022). Demand forecasting based machine learning algorithms on customer information: An applied approach. *International Journal of Information Management, 14*(3). https://doi.org/10.1007/s41870-022-00875-3

Mayer, A., Chardonnet, J., Häfner, P., & Ovtcharova, J. (2023). *Collaborative Work Enabled by Immersive Environments*. Springer. https://doi.org/10.1007/978-3-031-26490-0_6

Mei, Y., Lim, M. K., Qu, Y., Ni, D., & Xiao, Z. (2023). Supply chain risk management with machine learning technology: A literature review and future research directions. *Computers & Industrial Engineering, 175*. https://doi.org/10.1016/j.cie.2022.108859

Mertens, J., Klikovits, S., Bordeleau, F., Denil, J., & Haugen, Ø. (2024). Continuous evolution of digital twins using the DarTwin notation. *Software and Systems Modeling, 24*, 1405–1426 https://doi.org/10.1007/s10270-024-01216-7

Mohsen, S., Roza, D., & Behrooz, A. (2023). AI-powered blockchain technology in Industry 4.0, a review. *Journal of Economy and Technology, 1*, 222–241. https://doi.org/10.1016/j.ject.2024.01.001

Mueen, U., Khaled, S., Raja, J., & Sasa, P. (2021). Blockchain for drug traceability: Architectures and open challenges. *Health Informatics Journal, 27*(2), 1–28. https://doi.org/10.1177/14604582211011228

Musamih, A., Yaqoob, I., Salah, K., Jayaraman, R., Al-Hammadi, Y., Omar, M., & Ellahham, S. (2022). Metaverse in healthcare: Applications, challenges, and future directions. *IEEE Consumer Electronics Magazine, 12*, 33–46. https://doi.org/10.1109/MCE.2022.3223522

Mystakidis, S. (2022). Metaverse. *Encyclopedia, 2*, 486–497. https://doi.org/10.3390/encyclopedia2010031

Painuly, S., Sharma, S., & Matta, P. (2023). Artificial Intelligence in E-Healthcare Supply Chain Management System: Challenges and Future Trends. In *2023 International Conference on Sustainable Computing and Data Communication Systems (ICSCDS)*, Erode, India.

Papadopoulos, P., Pitropakis, N., & Buchanan, W. J. (2022). Decentralized privacy: A distributed ledger approach. In C. M. Hussain & P. Di Sia (Eds.), *Handbook of Smart Materials, Technologies, and Devices*. Springer. https://doi.org/10.1007/978-3-030-84205-5_58

Pedro, S., Augusto, D. C. R., Lino, M., & Julio, C. F. G. (2023). The influence of supply chain risk management in healthcare supply chains performance. *Production Planning & Control, 35*, 1–16. https://doi.org/10.1080/09537287.2023.2182726

Peukert, C., Weinhardt, C., Hinz, O., & Van Der Aalst, W. M. (2022). Metaverse: How to approach its challenges from a BISE perspective. *Business and Information Systems Engineering*, *64*(4), 401–406. https://doi.org/10.1007/s12599-022-00765-9

Phillip, W., Indira, K. D., Hisham, D., & Magdy, B. (2022). A survey on security in internet of things with a focus on the impact of emerging technologies. *Internet of Things*, *19*. https://doi.org/10.1016/j.iot.2022.100564

Qvarfordt, M., & Lagrosen, S. (2024). Healthcare digitalisation and its association with quality and employee health, a mixed-methods study. *International Journal of Workplace Health Management*, *17*(1), 1–20. https://doi.org/10.1108/IJWHM-03-2022-0045

Reddy, J., Elsayed, N., Elsayed, Z., & Ozer, M. (2023). A review on data breaches in healthcare security systems. *International Journal of Computers and Applications*, *184*(45), 1–7.

Rejeb, A., Rejeb, K., Treiblmaier, H., Appolloni, A., Alghamdi, S., Alhasawi, Y., & Iranmanesh, M. (2023). The Internet of Things (IoT) in healthcare: Taking stock and moving forward. *Internet of Things*, *22*. https://doi.org/10.1016/j.iot.2023.100721

Renu, N. (2021). Applications of AR and VR technologies in healthcare marketing. *Journal of Marketing Management*, *9*(2), 35–39.

Reza, T., Vahid, S., Ali, N., Pejvak, O., & Maria, F. (2021). Artificial intelligence in supply chain management: A systematic literature review. *Journal of Business Research*, *122*, 502–517. https://doi.org/10.1016/j.jbusres.2020.09.009

Santus, E., Marino, N., Cirillo, D., Chersoni, E., Montagud, A., Santuccione Chadha, A., & Lindvall, C. (2021). Artificial intelligence–aided precision medicine for COVID-19: Strategic areas of research and development. *Journal of Medical Internet Research*, *23*(3). https://doi.org/10.2196/22453

Sarode, R., Poudel, M., Shrestha, S., & Bhalla, S. (2021). Blockchain for committing peer-to-peer transactions using distributed ledger technologies. *International Journal of Computational Science*, *24*(3), 215. https://doi.org/10.1504/IJCSE.2021.115651

Saroj, K. N., Panda, S. K., & Madhabananda, D. (2023). Medical supply chain integrated with blockchain and IoT to track the logistics of medical products. *Multimedia Tools and Applications*, *82*, 32917–32939. https://doi.org/10.1007/s11042-023-14846-8

Sebastian, H., Tobias, R., Eric, C., Markus, B., & Helmut, K. (2020). The digital transformation of the healthcare industry: Exploring the rise of emerging platform ecosystems and their influence on the role of patients. *BuR – Business Research*, *13*(4), 1–37. https://doi.org/10.1007/s40685-020-00125-x

Seckin, D., & Koksalmis, G. H. (2024). A bibliometric analysis of metaverse technologies in healthcare services. *Service Business*, *18*, 223–254. https://doi.org/10.1007/s11628-024-00553-3

Senbekov, M., Saliev, T., Bukeyeva, Z., Almabayeva, A., Zhanaliyeva, M., Aitenova, N., & Fakhradiyev, I. (2020). The recent progress and applications of digital technologies in healthcare: A review. *International Journal Telemed Application*. https://doi.org/10.1155/2020/8830200

Senthil Kumar, J., & Mohamed, R. R. (2022). Building digital twins of cyber physical systems with metaverse for Industry 5.0 and beyond. *IT Professional*, *24*(6), 34–40. https://doi.org/10.1109/MITP.2022.3225064

Seyedan, M., & Mafakheri, F. (2020). Predictive big data analytics for supply chain demand forecasting: Methods, applications, and research opportunities. *Journal of Big Data*, *7*(53). https://doi.org/10.1186/s40537-020-00329-2

Shin, S., Koo, C., Kim, J., & Gursoy, D. (2024). Effects of metaverse experience on behavioral intention of visitors: Moderating role of similarity between virtual and real experience. *International Journal of Contemporary Hospitality Management*, *36*(12), 4055–4073. https://doi.org/10.1108/IJCHM-10-2023-1567

Shree, D. N., Krishna, D. V. L., & Patan, R. (2022). Securing Healthcare Data Using Decentralized Approach. In *2022 International Conference on Electronics and Renewable Systems (ICEARS), Erode, India.*, 1161–1165. DOI: 10.1109/ICEARS53579.2022.9752114

Siful, I., & Kutub, U. A. (2024). Decentralized vs. centralized database solutions in block chain: Advantages, challenges, and use cases. *Global Mainstream Journal of Innovation Engineering & Emerging Technology, 3*(4), 58–68. https://doi.org/10.62304/jieet.v3i04.195

Sodhro, A. H., Pirbhulal, S., & Sangaiah, A. K. (2018). Convergence of IoT and product lifecycle management in medical health care. *Future Generation Computer Systems, 86*, 380–391. https://doi.org/10.1016/j.future.2018.03.052

Srilakshmi, C., Uma, K., Rakesh, K. Y., Daya, S. K. V., Dharani, N. P., & Priyanka, S. (2024). IoT sensor data retrieval and analysis through cloud environment for effective power management. *Measurement: Sensors, 31.* https://doi.org/10.1016/j.measen.2023.100994

Stephenson, N. (1994). *Snow Crash.* Penguin Books Limited. https://books.google.co.in/books?id=inYs79gV4UQC

Suali, A. S., Srai, J. S., & Tsolakis, N. (2024). The role of digital platforms in e-commerce food supply chain resilience under exogenous disruptions. *Supply Chain Management, 29*(3), 573–601. https://doi.org/10.1108/SCM-02-2023-0064

Sullivan, C., & Staib, A. (2018). Digital disruption 'syndromes' in a hospital: Important considerations for the quality and safety of patient care during rapid digital transformation. *Australian Health Review, 42*(3), 294–298. https://doi.org/10.1071/AH16294

Tarun, K. V., Vikas, S., Kewal, K. S., Bhupendra, K., Sachin, C., & Rajneesh, P. (2023). Virtual reality (VR) and augmented reality (AR) transforming medical applications. In A. Khang (Ed.), *AI and IoT-Based Technologies for Precision Medicine* (pp. 324–348). IGI Global Scientific Publishing InfoSci. https://doi.org/10.4018/979-8-3693-0876-9.ch020

Thomas, Y. C., Julie, J. L., Dale, S. R., Tobias, S., & Stephan, M. W. (2020). *The Oxford Handbook of Supply Chain Management. Oxford Academic*, https://doi.org/10.1093/oxfordhb/9780190066727.001.0001

Tugce, B., Aziz Kemal, K., & Akdag, H. C. (2019). *Lecture Notes in Mechanical Engineering.* Springer. https://doi.org/10.1007/978-3-030-31343-2_50

Ucar, A., Karakose, M., & Kırımça, N. (2024). Artificial intelligence for predictive maintenance applications: Key components, trustworthiness, and future trend. *Applied Sciences, 14*(2). https://doi.org/10.3390/app14020898

Uddin, M., Obaidat, M., Manickam, S., Laghari, S. U. A., Dandoush, A., Ullah, H., & Ullah, S. S. (2024). Exploring the convergence of metaverse, blockchain, and AI: A comprehensive survey of enabling technologies, applications, challenges, and future directions. *Wiley Interdisciplinary Reviews: Data Mining and Knowledge Discovery, 14*(6), 1556. https://doi.org/10.1002/widm.1556

Vijaykumar, V., Mercy, P., Lucia Agnes Beena, T., Leena, H. M., & Savarimuthu, C. (2024). Convergence of IoT, artificial intelligence and blockchain approaches for supply chain management. In D. V. Grover, D. B. B. Balusamy, D. M. Milanova, & D. A. Y. Felix (Eds.), *Blockchain, IoT, and AI Technologies for Supply Chain Management.* Apress. https://doi.org/10.1007/979-8-8688-0315-4_2

Wang, Y., Su, Z., Zhang, N., Xing, R., Liu, D., Luan, T. H., & Shen, X. (2022). A survey on metaverse: Fundamentals, security, and privacy. *IEEE Communications Surveys & Tutorials, 25*(1), 319–352. https://doi.org/10.36227/techrxiv.19255058.v1

Wang, Z.-J., Chen, Z.-S., Xiao, L., Su, Q., Govindan, K., & Skibniewski, M. J. (2023). Blockchain adoption in sustainable supply chains for Industry 5.0: A multistakeholder perspective. *Journal of Innovation & Knowledge, 8*(4). https://doi.org/10.1016/j.jik.2023.100425

Wiederhold, B. K. (2022). Ready (or not) player one: Initial musings on the metaverse. *Cyberpsychology, Behavior, and Social Networking*, *25*(1), 1–2. https://doi.org/10.1080/00207543.2023.2240900

Wu, P., Chen, D., & Zhang, R. (2024). Topic prevalence and trends of metaverse in healthcare: A bibliometric analysis. *Data Science and Management*, *7*(2), 129–143. https://doi.org/10.1016/j.dsm.2023.12.003

Wolf, W. (2009). Cyber-physical systems. *Computer*, *42*(3), 88–89. 10.1109/MC.2009.81

Yan, C. M. Z., & Chen, C. W. (2019). Scalable access control for privacy-aware media sharing. *IEEE Transactions on Multimedia*, *21*(1), 173–183. https://doi.org/10.1109/COMST.2022.3202047

Yu, L., Shenle, P., & Eric, B. (2024). Unveiling the potential of digital twins in logistics and supply chain management: Services, capabilities, and research opportunities. *Digital Engineering*, *3*(100025). https://doi.org/10.1016/j.dte.2024.100025

Zaman, U., Mehmood, F., Iqbal, N., Kim, J., & Ibrahim, M. T. (2022). Towards secure and intelligent Internet of Health Things: A survey of enabling technologies and applications. *Electronics*, *11*(12), 1893. https://doi.org/10.3390/electronics11121893

Zeeshan, A., Asokan, V., Shad, A. K., Mudassar, M., Yusri, Y., & Devarajanayaka, M. (2024a). Beyond the screen: Ethical dilemmas in the XR frontier. In K. Tanveer, K. Pooja, & S. Nripendra (Eds.), *Multidisciplinary Applications of Extended Reality for Human Experience* (p. 29). IGI Global Scientific Publishing InfoSci. https://doi.org/10.4018/979-8-3693-2432-5.ch002

Zeeshan, A., Shahryar, S., Waqas, S., Mudassar, M., Yusri, Y., Surriyya, S., & Asokan, V. (2024b). Navigating the educational frontier and unveiling the role of virtual reality in faculty development. In A. Naim, A. Saklani, S. Khan, & P. Malik (Eds.), *An Emerging Consideration in Global Accreditation*. IGI Global Scientific Publishing InfoSci. https://doi.org/10.4018/979-8-3693-1698-6.ch015

Zhen-Song, C., & Jie-Qun, R. (2024). Metaverse healthcare supply chain: Conceptual framework and barrier identification. *Engineering Applications of Artificial Intelligence*, *133*. https://doi.org/10.1016/j.engappai.2024.108113

Part IV

Circular Economy and Innovation

9 Circular Economy and Technological Innovation Practices in Asia

Shahzad Ahmad Khan, Ibrahim Rashid Al Shamsi, Shahid Iqbal, and Adnan Ahmed Sheikh

9.1 INTRODUCTION

The circular economy (CE) offers sustainable development through designed systems which use regenerative processes yet work toward dissociating economic growth from resources uses (Ellen MacArthur Foundation (EMF), 2013). The CE works against traditional linear economic systems by prioritizing three core principles which contain both product lifecycle extension and material reuse practices together with waste-diminution techniques. Such principles help minimize damage to the environment while effectively using existing resources to create programs that promote strong environments as well as competitive business landscapes. The CE framework has been implemented by China together with Japan, South Korea, India, along with Singapore as a strategy that enhances their economic competitiveness and sustainability and addresses mounting waste and resource dilemmas (Ghisellini et al., 2016; Yuan et al., 2006). Governments across these countries understand the linear economy model becomes impractical for ongoing longevity because of growing population numbers along with limited resources along with concerns about climate change.

9.1.1 Circular Economy and the Asian Context

Industrial boom in Asian countries over the past few decades has fueled unprecedented economic growth, lifting millions out of poverty and transforming urban infrastructure. However, this progress has come at a cost: environmental degradation, resource depletion, and rising waste levels. According to the United Nations Economic and Social Commission for Asia and the Pacific (UNESCAP, 2021), Asia-Pacific generates over half of the world's solid waste, much of which ends up in landfills or oceans. Consequently, CE principles are increasingly seen as essential to reconciling economic development with environmental stewardship.

Governments across Asia are now taking deliberate steps to shift from a consumption-driven model to a more regenerative one. In China, for example, the concept of the CE was formally introduced in the early 2000s and was later institutionalized through the Circular Economy Promotion Law of 2009. This policy framework promotes eco-industrial parks (EIPs), waste-to-energy systems, and green

DOI: 10.1201/9781003611905-13

manufacturing practices (Zhu et al., 2010). The Chinese model emphasizes a top-down, state-led approach to CE implementation, integrating environmental objectives into national five-year development plans.

Similarly, Japan has long been a frontrunner in resource efficiency, leveraging its Sound Material-Cycle Society initiative, which seeks to reduce natural resource input and increase the reuse and recycling of materials. As part of this strategy, Japan employs extended producer responsibility (EPR) schemes and has robust systems for managing electronic and household waste (Hotta et al., 2008). The integration of high-end technologies such as smart sorting systems and artificial intelligence (AI)-driven material recovery enhances these efforts.

9.2 EMERGING TRENDS ACROSS ASIA IN THE CIRCULAR ECONOMY

As the global economy transitions toward sustainable and regenerative practices, Asia is rapidly emerging as a critical driver of CE transformation. Home to more than 60% of the world's population and some of the fastest-growing economies, the region faces immense challenges related to resource depletion, waste generation, and environmental degradation. In response, Asian countries are not only adopting circular principles but are also innovating unique models and forging new policy directions that align with their economic realities, technological capabilities, and societal needs. This chapter explores key emerging trends shaping the trajectory of the CE across Asia.

9.2.1 Policy-Driven Circularity

Governments across Asia are increasingly embedding circular economy concepts into national development plans, environmental policies, and industrial strategies. A standout example is China, which introduced CE principles as part of its 11th Five-Year Plan and later codified them through the Circular Economy Promotion Law (2009). This legal framework mandates industrial symbiosis, waste minimization, and cleaner production in key sectors (Yuan et al., 2006). China's model represents a top-down approach, with strong government intervention guiding CE implementation through fiscal incentives, monitoring systems, and performance metrics.

In contrast, Japan has institutionalized circularity through its Sound Material-Cycle Society framework, with clear targets for resource productivity, waste reduction, and recycling rates. Supported by public-private partnerships, Japan's approach includes EPR, green procurement policies, and consumer awareness campaigns (Hotta et al., 2008). Countries such as Indonesia, Vietnam, and India are also developing national CE roadmaps with support from international agencies such as UNEP (United Nations Environment Program) and the Ellen MacArthur Foundation.

9.2.2 Digital Platforms Enabling Circularity

A defining trend in Asia is the use of digital platforms and mobile applications to facilitate circular transactions and consumer behavior. E-commerce giants such as Alibaba (China) and Lazada (Southeast Asia) have introduced features to promote second-hand goods resale, product rental, and green logistics. Alibaba's platform Xianyu (Idle Fish) is one of the largest peer-to-peer resale markets in the world, boasting over 300 million users and contributing significantly to product reuse and waste reduction (Chen & Zhang, 2021).

Similarly, platforms such as Carousell (Singapore) and OLX India promote circular consumption by enabling consumers to buy and sell used items efficiently. These apps have lowered barriers to participation in the CE and have helped normalize reuse and recycling, especially among younger, tech-savvy populations.

9.2.3 Industrial Symbiosis and Eco-Industrial Parks

Asia is witnessing rapid growth in EIPs that promote industrial symbiosis, where waste or by-products from one company are used as inputs by another. South Korea's Ulsan EIP is often cited as a leading model where 137 companies collaborate to share resources, water, and energy, resulting in reduced emissions and operational costs (Park et al., 2008).

In China, more than 50 national EIPs have been developed under the guidance of the Ministry of Ecology and Environment, with pilot projects supported by UNIDO (United Nations Industrial Development Organization). These parks demonstrate how collaborative, region-based industrial models can foster circularity and stimulate local economies, particularly in resource-intensive sectors like chemicals, metallurgy, and textiles.

9.2.4 Bio-Based and Circular Materials Innovation

Another emerging trend is the development of bio-based materials and circular product design. In Thailand, startups like EnerGaia are producing biodegradable packaging from spirulina, while in India, companies such as Bakeys and Ecoware offer edible cutlery and compostable containers as alternatives to single-use plastic. These innovations are especially significant in light of Asia's mounting plastic waste crisis—countries like Indonesia, the Philippines, and Vietnam are among the top contributors to marine plastic leakage (Jambeck et al., 2015).

Singapore has launched the Sustainable Packaging Agreement and supported R&D into circular packaging through its Zero Waste Masterplan. The Asian Development Bank (2022) reports that innovation in bio-materials is increasingly backed by government funds and startup ecosystems, especially in ASEAN member states.

9.2.5 Circularity in the Built Environment

In response to urbanization and resource scarcity, Asia is integrating CE principles into urban planning and the built environment. In Hong Kong, the Construction Industry Council has launched initiatives promoting the use of recycled construction materials and the adoption of modular integrated construction techniques to reduce construction waste and carbon emissions.

Rashid et al. (2013) describe circularity in business models and supply chains as a precondition for sustainable manufacturing, which is necessary for the improved economic and environmental performance of industrialized and developing countries. Similarly, Läpple (2007) describes a CE as an important element of sustainable development. A much stronger conditional relationship is assumed by the Ellen MacArthur Foundation (EMF) (2013) and Webster (2015). Maybe even more pronounced, at least in the environmental dimension, are Bakker et al. (2014), who consider circularity as necessary for sustaining economic output. A similar approach is also held by the United Nations Environment Programme (2006) Bauer, S. (2009), which presents the CE as an essential condition for maintaining economic growth in a sustainable way, but here other pathways for establishing this condition are not excluded.

Singapore's Housing Development Board recycles concrete debris into new construction materials for public housing projects, while India's GIZ-backed (Deutsche Gesellschaft für Internationale Zusammenarbeit), C&D (Construction and Demolition) Waste Guidelines are promoting the recycling of construction and demolition debris into pavement blocks and road sub-bases (Witjes & Lozano, 2016). These trends highlight how urban infrastructure is being reimagined through the lens of circularity.

9.2.6 Sustainable Finance and Investment in CE Startups

Finally, there is increasing interest in sustainable finance mechanisms that support circular initiatives. The Asian Infrastructure Investment Bank and Asian Development Bank have launched green financing programs that include CE criteria. In South Korea, the government-backed Korea Development Bank has introduced green bonds for startups working in recycling, clean manufacturing, and eco-design.

Venture capital investment in CE-focused startups is growing across Asia, particularly in India, Indonesia, and Malaysia, where waste-tech and circular fashion are attracting attention. This trend reflects the alignment of environmental impact with financial viability, pushing circular innovation into mainstream entrepreneurship.

9.3 ROLE OF INNOVATION AND TECHNOLOGY

Technological evolution in partnership with innovation functions as an industrial and socio-economic transformation mechanism which drives significant changes throughout all markets and civil systems. Multiple scholars from academia have shown how these complementary elements trigger change while creating fresh approaches that consistently deliver superior user experiences. Napolitano et al. (2017) delve into the concept of autocatalytic networks within technological fields. The researchers demonstrate how several technological systems enhance the development of others through interconnected relationships which promote faster innovation rates. The system-like connection between different technological developments creates an environment where each progress in one area speeds up advancements across all other areas. According to He et al. (2024) exaptation represents a new concept in technological evolution through which existing technologies are reassigned for fresh purposes. Breakthrough innovations result from intersystem knowledge transfer according to their research which showcases that applying already developed solutions to fresh problems creates powerful effects. Interdisciplinary approaches play a vital role in advancing technology as they demonstrate the necessity of thinking from different angles during the development process.

9.3.1 Digital Technologies: The Backbone of Circular Systems

One of the foundational elements enabling circular transformation is the digitalization of supply chains and resource management systems. Technologies such as the Internet of Things (IoT), cloud computing, and AI allow real-time monitoring of product life cycles, energy flows, and material usage. According to Ellen MacArthur Foundation (2019), digital tools help businesses design waste out of systems, keep materials in use for longer, and regenerate natural ecosystems.

For instance, IoT-based sensors installed in smart bins can measure waste volumes and trigger efficient collection schedules, reducing emissions and labor costs. AI algorithms are applied in waste sorting facilities to identify and separate recyclable materials with high precision, significantly improving recycling rates and reducing contamination (Zhou et al., 2021). These applications contribute to greater operational efficiency and facilitate closed-loop resource systems, especially in sectors like manufacturing and urban logistics.

9.3.2 Blockchain: Ensuring Transparency and Traceability

Blockchain technology is also gaining prominence in CE implementation due to its ability to ensure traceability, transparency, and trust across supply chains. By creating immutable digital ledgers, blockchain allows stakeholders to verify the origin, usage, and recycling history of materials and products.

In the textile industry, blockchain is used to track the environmental and ethical credentials of fabrics from raw material sourcing to retail. Platforms like Circularis and Provenance have created blockchain-based solutions that authenticate the sustainability claims of products (De Man & Strandhagen, 2017). Similarly, in China, Alibaba's logistics arm Cainiao, has piloted blockchain-enabled traceability to monitor green packaging use and recyclable logistics materials, demonstrating the potential for wide-scale application in circular logistics.

9.3.3 Advanced Materials and Product Design

Beyond digital tools, innovation in materials science and product design plays a crucial role in enabling circularity. The development of biodegradable polymers, modular components, and easily separable assemblies empowers industries to rethink how products are designed, used, and disposed of. According to Bocken et al. (2016), circular innovation often begins at the design stage, where decisions regarding durability, disassembly, and material selection can predetermine a product's environmental impact.

In Asia, Singapore-based startups such as Alterpacks and Magorium are creating biodegradable packaging from food waste and chemically recycled plastics, respectively. These innovations reduce reliance on virgin fossil-based materials and support Singapore's Zero Waste Masterplan (2019). Likewise, Japan's Hitachi is investing in designing modular appliances that can be upgraded or repaired rather than replaced, aligning with product longevity goals and reducing material throughput.

9.3.4 Smart Circular Cities and Urban Innovation

Technological innovation is also transforming how cities address urban circularity. Smart cities leverage integrated systems—such as real-time energy monitoring, water recycling, and automated waste management—to reduce environmental footprints. In South Korea, Songdo Smart City has integrated pneumatic waste systems that transport waste via underground pipes to central sorting centers, eliminating the need for traditional collection methods and increasing waste recycling efficiency (Kim, 2020).

In India, the city of Indore—a model for smart city development—uses GPS-enabled waste trucks, AI-driven sanitation monitoring, and citizen apps for waste complaints. These technologies not only foster resource optimization but also engage citizens in circular behavior, an often-overlooked aspect of CE implementation.

9.4 EMERGING BUSINESS MODELS ENABLED BY TECHNOLOGY

Technology has also unlocked new business models that facilitate circularity. The rise of product-as-a-service, shared economy platforms, and digital resale marketplaces reflects the shifting focus from ownership to access and reuse. For example, electronics companies like Fairphone and Dell offer modular, serviceable devices while retaining responsibility for take-back and refurbishment, reducing e-waste and encouraging reuse. Asian platforms such as Xianyu (Idle Fish) by Alibaba and Carousell in Singapore have created vibrant marketplaces for second-hand goods, reducing the consumption of new products and promoting reuse. According to Lacy and Rutqvist (2015), such digitally enabled circular models not only generate revenue but also cultivate long-term customer relationships based on sustainability values. According to He et al. (2024) exaptation represents a new concept in technological evolution through which existing technologies are reassigned for fresh purposes. Specifically, the implementation of blockchain technology can greatly improve the CE by facilitating better transparency, traceability, and efficiency within the supply chains. Addressing the challenges of implementing CE principles like tracking product life cycle, recovering resources, and encouraging sustainable consumption poses the use of decentralization and blockchain's immutable framework. Blockchain allows for complete traceability of materials and products during their life cycles. A shared ledger system allows for the recording of all events in the supply chain from the procuring of raw materials to disposal of a product. Centobelli et al. (2021) observes that the lack of transparency and traceability is one of the most important barriers to circular supply chains while proposing blockchain technology for incorporating recycling and remanufacturing systems. Blockchain supports the tracking of products and materials in real-time throughout their lifecycle which ensures authenticity as well as eases recycling and reuse processes. Stakeholders are able to trust the business and its governance framework due to the transparency. Smart contracts take manual work out of transaction completion and agreement enforcement without intermediaries; this simplifies product return, recycling, and resource sharing." Circular business models experience lower expenses and enhanced effectiveness because of these automations. In addition, blockchain adoption requires technological maturity and knowledge sharing, and blockchain-based solutions will significantly enhance competitive performance and supply chain innovation capabilities (Alkathiri & Mudhsh, 2024). The distributed nature of blockchain tackles supply chain challenges by improving traceability and sharing of information; leading to the competitive advantage (Dong et al., 2024). Supply chain-based solutions will also enhance supply chain resilience and agility; helping to support performance in uncertain environments (Sudrajat et al., 2023).

9.5 REAL-WORLD APPLICATIONS AND INDUSTRY TRANSFORMATION

The "Real-World Applications and Industry Transformation" section, providing more examples, regional insights, and scholarly depth to showcase how various industries in Asia are integrating CE principles.

9.5.1 Real-World Applications and Industry Transformation in the Circular Economy Across Asia

As CE gains global traction, industries in Asia are beginning to adopt and implement principles to redefine production systems, reduce waste, and drive innovation. The transformation is particularly evident in sectors such as manufacturing, textiles, construction, electronics, and food. These industries, traditionally associated with high levels of resource consumption and environmental impact, are now at the forefront of reimagining their operations to align with circularity.

9.5.2 South Korea: Industrial Symbiosis in Heavy Manufacturing

South Korea has emerged as a leader in integrating CE practices within its industrial sector. One of the most cited examples is POSCO, a global steel manufacturer headquartered in Pohang. POSCO has implemented a comprehensive by-product synergy strategy, where waste materials generated during steel production—such as slag, dust, and sludge—are repurposed for use in other industrial processes.

For instance, steel slag is processed and used as raw material in cement production, road base construction, and agriculture, thus preventing it from being landfilled. According to the Korea Environmental Industry & Technology Institute, such symbiotic initiatives have led to significant cost savings and waste reduction while simultaneously generating revenue through by-product sales.

Moreover, South Korea's EIPs, supported by the Ministry of Environment and UNIDO, facilitate collaboration among different manufacturing firms within a geographic cluster. These parks promote the exchange of waste materials, utilities, and services to optimize resource usage—a model that is now being replicated in several Asian countries.

9.5.3 India: Circular Models in Textiles and Electronics

India's booming textile and apparel industry, one of the largest globally, is also beginning to embrace circularity, driven by environmental regulation, export market pressures, and local innovation. Companies like Arvind Limited and Aditya Birla Fashion are investing in circular fabrics, including organic cotton, recycled polyester, and eco-friendly dyes. These materials are designed to reduce environmental impact across the product life cycle.

Additionally, the rise of textile-to-textile recycling in India is notable. Startups such as Resha Mandi and Chindi collect post-consumer textile waste, process it using mechanical and chemical recycling methods, and reintroduce it into the manufacturing cycle. This approach addresses the growing concern around textile waste, especially in urban centers like Mumbai and Bengaluru.

In the electronics sector, the issue of e-waste has become increasingly urgent. India ranks as the third-largest generator of e-waste globally. Circular initiatives such as Attero Recycling and E-Parisaraa have pioneered processes to recover valuable metals (gold, copper, rare earth elements) from discarded smartphones, laptops, and circuit boards. These companies also provide collection, dismantling, and material recovery services to OEMs (Original Equipment Manufacturers) under the EPR framework introduced by the Indian government.

9.5.4 China: From Waste Crisis to Circular Innovation

China's CE efforts are deeply embedded in policy, as evidenced by the Circular Economy Promotion Law of 2009 and subsequent action plans. Chinese industries are encouraged to reduce, reuse, and recycle resources through tax incentives, green certifications, and compliance targets.

A prime example of transformation comes from the construction sector, which contributes to nearly 40% of China's solid waste. Companies like China State Construction Engineering Corporation now utilize recycled aggregates from demolished buildings to produce concrete for new infrastructure projects. This not only reduces the extraction of virgin materials but also addresses the problem of construction waste in rapidly urbanizing cities.

In consumer goods, Alibaba Group has introduced a CE e-commerce platform called Idle Fish (Xianyu), where users can buy, sell, or exchange used items, promoting reuse and reducing landfill pressure. With millions of users, Idle Fish exemplifies how digital platforms can support peer-to-peer circular models at scale.

9.5.5 Singapore: Smart Circular Economy through Technology

Singapore's compact geography and limited natural resources have made it a pioneer in urban circularity. Under the Zero Waste Masterplan (2019), Singapore targets 70%

recycling rates by 2030. The city-state has developed smart waste systems that include IoT-enabled recycling bins, AI-powered material recovery facilities, and automated pneumatic waste collection systems in high-density housing developments.

In the food sector, companies like UglyGood and Treatsure use surplus food from restaurants and supermarkets to produce animal feed, skincare products, and discounted meal boxes. These business models are supported by government grants and research partnerships, creating a circular ecosystem that diverts organic waste from incineration.

9.6 CONCLUSION

The adoption of circular economic principles is driving significant industry transformation across Asia. While approaches vary by country and sector, the common thread is the integration of circularity into core business strategies, supported by innovation, regulation, and collaboration. Real-world applications—ranging from steel recycling in South Korea to e-waste recovery in India and digital resale in China—illustrate that circular practices are not only environmentally essential but also economically viable. Wide-reaching CE solutions emanate from local innovators who receive backing from digital platforms and economic market drivers to make substantial progress even in developing economic regions (Chaturvedi et al., 2021).

The CE is no longer a conceptual aspiration but a practical necessity, especially for Asia—a region grappling with the dual pressures of environmental degradation and economic growth. Encouragingly, several Asian countries have started to embrace CE principles by integrating policy frameworks, technological innovation, and stakeholder collaboration. From top-down government initiatives to grassroots innovation, the region presents a diverse and evolving landscape of circular practices. As these efforts continue to mature, they hold the potential to position Asia not just as a global manufacturing hub, but as a leader in sustainable industrial transformation. Asia's shift toward a CE, driven by technological innovation, offers a sustainable pathway for the region's future growth. unlocking new economic opportunities. However, continued investment in technology, supportive policies, and cross-sector collaboration will be essential. As Asia continues to embrace these innovations, it can emerge as a global leader in the transition to a more sustainable, resource-efficient future.

REFERENCES

Alkathiri, A. A. A., & Mudhsh, B. A. (2024). Analysis of tense system in Jibbali (Shehri) language. *Pakistan Journal of Life and Social Sciences*, *22*(2), 1219–1230.

Asian Development Bank. (2022). *Towards a Circular Economy in Asia*. ADB Publications.

Bakker, C. A., Hollander, M. C., Hinte, E., & Zljlstra, Y. (2014). *Products That Last – Product Design for Circular Business Models*. TU Delft Library.

Bauer, S. (2009). The secretariat of the United Nations Environment Programme: Tangled up in blue. In *Managers of Global Change: The Influence of International Environmental Bureaucracies* (pp. 169–201). MIT Press.

Bocken, N. M., Bakker, C., & Pauw, I. D. (2016). Product design and business model strategies for a circular economy. *Journal of Industrial and Production Engineering*, *33*(5), 308–320.

Centobelli, P., Cerchione, R., Del Vecchio, P., Oropallo, E., & Secundo, G. (2021/2022). Blockchain technology for bridging trust, traceability and transparency in circular supply chains. *Information & Management*, *59*(7), 103508. https://doi.org/10.1016/j.im.2021.103508

Chaturvedi, A., Arora, R., & Sharma, M. (2021). Circular economy and sustainable consumption: Insights from Indian e-waste sector. *Resources, Conservation & Recycling*, *164*. https://doi.org/10.1108/IJPPM-06-2022-0314

Chen, Y., & Zhang, R. (2021). Digital platforms and circular economy: Case study of Idle Fish. *Journal of Cleaner Production*, *280*, 124478.

De Man, R., & Strandhagen, J. O. (2017). Blockchain technology for improving traceability in sustainable supply chains. *International Journal of Information Systems and Supply Chain Management*, *10*(1), 77–97.

Dong, X., Li, J. M., Lu, X. L., Lin, X. Y., Hong, M. Z., Weng, S., & Pan, J. S. (2024). Global burden of adult non-alcoholic fatty liver disease (NAFLD) and non-alcoholic steatohepatitis (NASH) has been steadily increasing over the past decades and is expected to persist in the future. *Translational Gastroenterology and Hepatology*, *9*, https://doi.org/10.21037/tgh-23-118.

Ghisellini, P., Cialani, C., & Ulgiati, S. (2016). A review on circular economy: The expected transition to a balanced interplay of environmental and economic systems. *Journal of Cleaner Production*, *114*, 11–32.

He, Y., Lee, J. D., & Lee, S. (2024). Innovation beyond intention: Harnessing exaptation for technological breakthroughs. arXiv preprint arXiv:2412.19662.

Hotta, Y., Hayashi, S., Bengtsson, M., & Mori, H. (2008). Extended producer responsibility in East Asia: Review of experiences. *International Review for Environmental Strategies*, *8*(1), 129–154.

Jambeck, J. R., Geyer, R., & Wilcox, C. et al (2015). Plastic waste inputs from land into the ocean. *Science*, *347*(6223), 768–771.

Kim, J. (2020). Smart cities for sustainable development: Case study of Songdo, South Korea. *Journal of Cleaner Production*, *247*, 119154.

Lacy, P., & Rutqvist, J. (2015). *Waste to Wealth: The Circular Economy Advantage*. Palgrave Macmillan.

Läpple, F., 2007. Abfall- und kreislaufwirtschaftlicher Transformations prozess in Deutschland und in China: Analyse, Vergleich, Übertragbarkeit. Universität Heidelberg, Heidelberg.

MacArthur, E., & Cowes, U. (2019). Artificial intelligence and the circular economy. *Ellen MacArthur Foundation, Cowes, UK*.

MacArthur, E. (2013). Towards the circular economy. *Journal of Industrial Ecology*, *2*(1), 23–44.

Napolitano, F., Napolitano, P., & Angelillo, I. F. (2017). Seasonal influenza vaccination in pregnant women: Knowledge, attitudes, and behaviors in Italy. *BMC Infectious Diseases*, *17*, 1–7.

Park, H. S., Rene, E. R., Chiu, A. S. F., & Yi, J. (2008). Strategies for sustainable development of industrial park in Ulsan, South Korea—From spontaneous evolution to systematic expansion of industrial symbiosis. *Journal of Environmental Management*, *87*(1), 1–13.

Rashid, A., Asif, F. M. A., Krajnik, P., & Nicolescu, C. M. (2013). Resource conservative manufacturing: An essential change in business and technology paradigm for sustainable manufacturing. *The Journal of Cleaner Production*, *57*, 166–177.

Standing Committee of the National People's Congress. (2008). *Circular Economy Promotion Law of the People's Republic of China* (promulgated 29 August 2008; effective 1 January 2009). Ministry of Ecology and Environment of the PRC.

Sudrajat, H., Rossetti, I., & Colmenares, J. C. (2023). Piezocatalysis: A promising alternative route for CO 2 reduction. *Journal of Materials Chemistry A*, *11*(45), 24566–24590.

United Nations Economic and Social Commission for Asia and the Pacific (UNESCAP, 2021) https://www.unescap.org/blog/circular-economy-making-sustainability-part-the-solution-asia-pacific#:~:text=The%20Asia%2DPacific%20region%2C%20hosting, supporting%20human%20health%20and%20livelihoods?

Webster, K., 2015. *The Circular Economy: A Wealth of Flows*. Ellen MacArthur Foundation.

Witjes, S., & Lozano, R. (2016). Towards a more circular economy: Proposing a framework linking sustainable public procurement and sustainable business models. *Resources, Conservation and Recycling*, *112*, 37–44.

Yuan, Z., Bi, J., & Moriguichi, Y. (2006). The circular economy: A new development strategy in China. *Journal of Industrial Ecology*, *10*(1–2), 4–8.

Zhou, Y., Zhang, C., & Wang, Q. (2021). Intelligent recycling and circular economy: The role of AI in improving material recovery. *Resources, Conservation and Recycling*, *170*, 105615.

Zhu, Q., Geng, Y., & Lai, K. H. (2010). Circular economy practices among Chinese manufacturers: Challenges and opportunities. *Journal of Environmental Management*, *91*(6), 1494–1501.

10 Toward a Sustainable Blue Economy through Sustainable Innovation

Opportunities and Challenges

Karima Sayari

10.1 INTRODUCTION

The blue economy is a set of economic activities associated with oceans, coasts, and islands. It encompasses a range of activities, including traditional industries such as shipping and fisheries, as well as emerging economic sectors, including renewable energy, biotechnology, and tourism. The blue economy has the potential to contribute to sustainable development and create jobs and growth.

Sustainable blue economy (SBE) plans can deliver noticeable socio-economic and environmental advantages for countries and regions. The blue economy presents an opportunity to ensure social equity, economic growth and ocean environmental sustainability (Wenhai et al., 2019). The ocean has a central role in global development and is the basis for the prosperity of human civilization. The core of the blue economy is the sustainable development of ocean resources, which can ensure long-term ocean environmental health and provide boundless ocean economic returns during the exploration and utilization of various ocean resources. The ocean covers nearly 71% of the Earth's surface and holds roughly 97% of the planet's water, accounting for the cradle of life on Earth. The ocean is one of the most important aspects of the Earth's environmental system. The ecosystem services provided by the ocean are vital for the survival and sustainable development of human society (Webb, 2023).

Oceans are increasingly viewed as a new frontier for exploration, development, and business opportunities. Numerous ocean-based industries are thriving around the world, including shipping and shipbuilding, fisheries and aquaculture, maritime tourism, offshore oil and gas exploration, ocean science research, and marine environmental protection. These industries are key contributors to the global economy, coastal nation economies, and national security. Economists have predicted that ocean-based industries will continue to grow robustly and could triple their global annual economic output to about $3 trillion in the next 10 to 20 years (United Nations Global Compact, 2020).

 DOI: 10.1201/9781003611905-14

Globally concern for oceans and coastal zones has been increasing. In 2012, the concept of "blue economy" was first put forward as part of a policy strategy to encourage nations to invest in ocean-related industries (Lee et al., 2020). Since then, numerous activities have been undertaken to explore and promote the blue economy at both international and local levels. In general, the blue economy presents attributes at the industrial level, aimed at promoting the sustainable use of ocean resources and the cooperative development of ocean industries.

However, the blue economy faces significant challenges, including climate change, overfishing, pollution, and habitat destruction. Political commitments have been made to a SBE at global, regional, and national levels. The OECD highlights sustainable innovation as a key enabler of the blue economy, drawing on existing technologies, innovation, and entrepreneurial activities in ocean-related sectors. However, crucial knowledge gaps currently exist regarding the state of sustainable innovation efforts, their opportunities, and their challenges across the blue economy sectors. This chapter aims to fill these gaps by assessing current research and cases on sustainable innovation in the blue economy.

10.1.1 Research Rationale and Research Gap

There is a growing interest and focus to research on blue economy agenda. The blue economy has gained significant interest over the past decade due to its focus on sustainable economic development in the ocean. However, there is still a lack of general understanding of what constitutes ocean sustainable development, particularly in developing countries. The academic literature on SBE is still limited despite the high political attention translated into conventions and treaties aiming at its sustainability (Lee et al., 2020; Techera & Winter, 2021; Wenhai et al., 2019), especially for integrating sustainable innovation to achieve a SBE. Scientific interest in SBE has spiked recently with a growing number of contributions published post-2020. An emerging research agenda is explored identifying the key concepts framing SBE studies, the thematic priorities guiding the current SBE research, the main academic journals disseminating SBE knowledge, and the leading institutions contributing to the SBE studies. With enormous potential for innovation and economic expansion, marine (blue) biotechnology is a new discipline that makes it possible to value new goods and procedures. This study aims to address the gap by analyzing previous research and literature while offering insights into the current state of sustainable innovation integration in blue economy navigating the opportunities and challenges.

Furthermore, the incorporation of innovation into blue economy presents significant implications for stakeholders, including regulators, policymakers, investors, financial institutions, and society at large, as it necessitates a comprehensive understanding of how sustainable innovation affects policy formulation, promotes innovation within the financial ecosystem, and assists investors in making critical investment decisions. Emphasizing the significance of SBE, sustainable innovation has emerged as a key driver, for developing a comprehensive perspective of blue economy.

Currently, most economists agree that the health and prosperity of humans depend on clean natural resources, such as the ocean and the marine industry. Seas and oceans are reliable providers of food, energy, minerals, health, and leisure (Yousef, 2024). The ocean and its resources need to be handled carefully to prevent endangering the riches and well-being of the whole world. A sustainable use of the oceans and seas is difficult to achieve unless all aspects of human activity that impact them are managed. Even when globally relevant measures are needed, it is very difficult to reverse the declining environmental trends brought on by unsustainable development practices at the local and regional levels. The marine industry and the sea are essential to human prosperity and well-being. Policies that aim to lessen the consequences of climate change, adapt to its aftereffects, and come up with novel solutions for problems like excessive consumption and the negative effects of economic activity must be put into place in order to overcome the barriers to sustainable development. Seas are a reliable source of food, energy, minerals, recreation, and health (Yasser et al., 2024).

10.1.2 Research Objectives

The primary aim of this chapter is to provide an in-depth examination of how sustainable innovation can drive the transition toward a SBE. This study aims to shed light on the avenues, opportunities, and challenges required to achieve a SBE through sustainable innovation. This will analyze the impact of innovation on the development of a SBE. Moreover, the chapter examines the potential and problems presented by this intersection of sustainable innovation and SBE.

The remaining chapter will be as follows: Section 10.2 analyses the scope of sustainable innovation in the blue economy. Section 10.3 runs through the opportunities and challenges of innovations in blue economy. Section 10.4 summarizes and concludes the chapter findings.

10.2 SUSTAINABLE INNOVATION IN THE BLUE ECONOMY

The emerging concept of a SBE focuses on maximizing the benefits of the ocean while prioritizing the sustainability of ocean ecosystems and resources. Sustainable innovation encompasses innovative products, processes, and practices that promote sustainable development by integrating environmental, social, and economic considerations in decision-making. Therefore, innovation in industries, practices, processes, and governance systems that augur ocean economic activities is essential for a SBE. To identify and explore the avenues, opportunities, and challenges in achieving a SBE through sustainable innovation, a general understanding of the SBE and sustainable innovation is provided first, before looking into the context of developing countries (Hemer et al., 2018).

Sustainable living requires societal transitions to environmental sustainability considerations. Innovation in goods and processes is vital to make this transition. Wide access to innovation augments society's ability to innovate (Wenhai et al., 2019). Selected academic literature shows how the blue economy considers the role of innovation in growth and development. The blue growth proposal stresses stimulating innovation in fisheries and aquaculture, biotechnology, maritime transport, coastal tourism, and ecosystems.

Understanding the trends and impending challenges impacting marine and coastal ecosystems is crucial, as is identifying opportunities for the development of innovative, sustainability-oriented solutions to address these concerns. Pace et al. (2023) revealed that alongside traditional decision support tools used in environmental management and conservation sciences, foresight methodologies provide comprehensive and detailed insights into potential long-term trends. In their study, they utilizes horizon scanning activity to examine emerging signals and patterns pertinent to future developments and innovations for the sustainable development of the blue economy, concentrating on the year 2030. It outlines many global transformations, challenges, and opportunities linked to the blue economy via an examination of significant international publications and studies. The results include a synthesis of 11 worldwide advancements in the blue economy and 7 overarching trends and forecasts. These may serve as a great resource for marine scientists in formulating a future research program or as evidence for decision-makers and policymakers addressing issues about the expansion of the blue economy.

10.3 OVERVIEW OF OPPORTUNITIES AND CHALLENGES

According to OECD (2024), ocean-based industries are projected to surpass overall global economic growth in terms of employment and value added by 2030. However, the seas have been deteriorating for years due to a variety of human activities, such as overfishing and the burning of fossil fuels. Global warming harms or kills marine plants, animals, and other creatures by increasing the absorption of carbon dioxide, which also lowers oxygen levels and acidifies the seas. Additionally, hundreds of millions of people living in coastal regions are becoming more and more vulnerable as sea levels rise as a result of the ice caps melting.

Through their findings Cisneros-Montemayor et al. (2022) discovered that the primary discrepancies in a region's potential to develop a blue economy are not attributable to the availability of natural resources, but rather to elements like infrastructure, national stability, and corruption, all of which may be enhanced by focused investments and cross-sector collaboration. To discover equitable routes for creating or altering ocean sectors, knowledge gaps may be addressed by combining historical natural and social scientific information on the causes and results of resource use and management.

Their findings imply that in order to support evidence-based, cooperative planning that guarantees that sectors are carefully selected, that local benefits are given priority, and that the blue economy achieves its social, environmental, and economic objectives, policymakers must include academics and stakeholders.

In order to promote ocean exploration and environmental monitoring, the Decade of Ocean Science supports efforts aimed at creating a more sustainable and integrated ocean observation network. The immense expanse of the ocean and its intricate coasts remain insufficiently explored and comprehended. The deep water represents a frontier in ocean sciences, where globally coordinated research teams consistently uncover new ocean phenomena of significant relevance, as well as novel creatures and chemicals with possible future applications. Continuous and methodical ocean observations are essential to record oceanic changes, initialize ocean system models, and provide vital information for enhanced comprehension of ocean dynamics (Visbeck, 2018).

10.4 OPPORTUNITIES FOR A SUSTAINABLE BLUE ECONOMY USING INNOVATIVE TECHNOLOGIES

According to OECD (2024), the ecosystem services provided by freshwater, coastal, and marine ecosystems include the provision of food and raw materials, water purification, carbon capture, biodiversity, and recreation.

Across the global ocean observing community, knowledge and experience have been compiled, focusing on what is being done to explore blue economy opportunities and what could be done to enhance exploration. Exploration examples from Australia, China, Europe, Latin America, and the United States have been gathered to illustrate a range of approaches that incorporate international perspectives. Although activity varies among countries and regions, and despite the blue economy being a relatively new concept, promising exploration has been demonstrated in a number of areas. Proposals for the development of the blue economy include shouldering global responsibilities to protect the marine ecological environment; strengthening international communication and sharing development achievements; and promoting the establishment of global blue partnerships. As a shared resource, the ocean supports all life on Earth. It provides food, water, energy, raw materials, and important ecological services, which are essential for human existence and sustainable development (Wenhai et al., 2019). The section focuses on transformation in a variety of areas when innovation is included into the blue economy.

10.4.1 Aquaculture Innovations: Technologies for Sustainable Fisheries and Aquaculture

Innovations that have been or are being developed, adopted and diffused in aquaculture include innovations relating to livestock breeding, feeding and feed formulation, production methods, management control, waste processing and environmental monitoring. Many of the innovations in aquaculture are technological innovations that are developed and diffused by external firms and research organizations rather than being internally developed by aquaculture enterprises. The development of new technologies and the adoption of such technologies and associated innovations adopted by aquaculture activities often occur in different countries.

In developing countries, prior to the adoption of a technological innovation there may need to be a number of supporting innovations that also need to be adopted. For example, the adoption of pelleted feeds in Asian shrimp aquaculture had to be preceded by the adoption of feed formulation and management control innovations needed to use the pelleted feeds effectively. Similarly, technology innovations that require more advanced environmental monitoring technologies to be adopted may be ineffective in countries where the management control innovations necessary for their effective use have not also been adopted (Stead, 2019).

10.4.2 Ocean Cleanup Technologies

The marine environment is under increasing pressure from pollution, over-exploitation of resources, and climate change. The European Green Deal addresses climate and environmental challenges, including a permanent Union ocean observation and data network, to ensure ocean sustainability. Innovative technologies are needed to achieve Blue Deal targets and protect, restore, and make the seas and oceans sustainable and resilient. Ocean plastic pollution is a growing challenge, and investment in new cleanup technologies is increasing. However, poorly designed or unregulated deployment of cleanup technologies can cause harm and further stress at-risk ecosystems. The goal of many new ocean cleanup technologies is to remove plastics before they cause harm to the environment, wildlife, and human health (Falk-Andersson et al., 2023). This is an urgent need, as plastics continue to be added to the oceans.

10.4.3 Renewable Energy Sources for Maritime Activities

The use of renewable energy sources in shared energy production systems is recognized as a key tactic to meet the pressing need to move toward a more sustainable society (Buonomano et al., 2024). In this regard, the Renewable Energy Communities (RECs) framework was developed by the European Directive 2018/2001 as a means of achieving the 2050 climate-neutrality objectives and aiding in the worldwide endeavor to combat climate change. Public and corporate organizations, as well as people, are urged to work together to promote the integration of clean technology in order to lower energy-related greenhouse gas emissions via the RECs' enabling framework.

The study of Nikolopoulou et al. (2019) suggests an innovative framework for ports to manage capacity, traffic, efficiency, and environmental issues in the future, assisted by disruptive technology. By reducing the overall environmental impact of intermodal connections, improving operational efficiency, increasing data sharing and information visibility, and promoting innovation in the port-urban context, the innovations to be implemented within the proposed framework will help the Port of the Future achieve its goals. The suggested solutions would support multimodality, container mobility and interconnection, digitalization, and the production of smart products.

Installing and producing marine renewable energy always has an influence on the environment, both positively and negatively. The location, the size and footprint of the installation, and the employment of certain technologies all affect these effects. The production of low-carbon power is one of the main benefits of ocean renewable energy. According to analysis, the carbon intensity of marine hydrokinetic resources like wave and tidal power and offshore wind is more than an order of magnitude lower than that of fossil fuel production (United Nations, 2016).

Amon et al. (2022) reveal that development of marine renewable energy, such as offshore wind, solar, and ocean energy, is a crucial component of the blue economy as it promotes energy security, a sustainable energy transition, and the reduction of CO_2 emissions that lessen the consequences of climate change.

10.4.4 Marine Biotechnology

Marine biotechnology involves the discovery, production, and use of biomolecules from the marine environment for the development of biopharmaceutical and other bio-based products (Rotter et al., 2023). Marine biota represent a largely undeveloped source of biomolecules with proven potential to be transformed into high-value products. For decades, marine biotechnology has been successfully used to develop new products for the pharmaceutical and cosmeceutical industries, but there is also interest in expanding the use of marine biota in food, feed, and food supplement products.

As such, marine biotechnology emerging value chains associated with the sustainable use of marine biota could contribute to a more SBE. A marine biotechnology value chain involves marine biota discovery and collection, bioactivity screening, target biomolecule and pharmacological compound identification, biomolecule retrieval and upscaling production, bioprocessing technology development, and product formulation and market introduction.

Producing organic materials with potential uses in the food and pharmaceutical industries is made feasible by biotechnology. The bioeconomy, which develops and uses renewable biological resources from land and water, includes this new and emerging industry (Martínez-Vázquez et al., 2021).

While there is considerable potential for growth, the blue economy is faced with a number of challenges, particularly around turning blue economy opportunities into genuine blue economy outcomes in a sustainable manner. The next section highlights key challenges that need to be addressed for blue economies to be both realized and sustainable.

10.5 CHALLENGES IN ACHIEVING A SUSTAINABLE BLUE ECONOMY

A number of obstacles impede the blue economy's ability to expand. Aquatic habitats have been seen and handled as boundless resources and essentially free garbage disposal sites for a vast portion of human history. But these resources are far from endless, and the effects of this strategy are becoming more and more apparent. The restricted coastal interface is oversubscribed by many businesses, and the effects of climate change are becoming more apparent. Rising demand, weak governance, insufficient financial incentives, technological developments, and a lack of management tools have all contributed to ineffectively regulated or unregulated user competition. As a result, coastal areas and valuable aquatic resources have been abused, sometimes causing irreversible harm. In this very competitive environment, the issues of the most dependent and vulnerable individuals—like small-scale artisanal fishermen—are often disregarded (Lee et al., 2020).

Although many nations strive to achieve a more environmentally friendly agenda by developing their maritime economy, these initiatives are still difficult to accomplish. Governments throughout the world must shift a little portion of their economies in the direction of a robust, global blue economy, a portion that goes toward funding R&D, education, modern infrastructure, technology, and job creation. It will be difficult to shift from an agricultural sector to a bluer economy. In order to make blue economies viable, governments must cooperate and exchange research and expertise.

10.5.1 Regulatory Barriers to Sustainable Practices

The blue economy is increasingly at the forefront of national and regional plans and policies for development. Rhetorically, the blue economy promises a rich future through the sustainable use of ocean resources from food security to renewable energy and biodiversity offsets. However, the surface glisten of the blue economy obscures the deeper challenges of securing a just and equitable blue economy for all.

Significant challenges remain within policy, governance, and management frameworks at all scales that may prevent emerging blue economy opportunities from being successfully and sustainably realized (Bax et al., 2022).

The 2030 Agenda for Sustainable Development and accompanying sustainable development goals (SDGs) present both opportunities and challenges for transforming ocean governance. The ocean is the focus of Goal 14: Conserve and sustainably use the oceans, seas and marine resources for sustainable development (United Nations, 2021).

10.5.2 Environmental Threats for Blue Economies

Based on the OECD Global Survey on Localizing the Blue Economy, we know climate change seriously threatens the blue economy, which causes great concern. While areas show more awareness for environmental issues (82%), cities and basins are most worried about the consequences of climate change. Since it generates half of Earth's oxygen, absorbs 25% of CO_2 emissions, and stores 93% of the heat from these emissions, the ocean is vital for controlling the global temperature.

Climate change clearly and extensively influences marine habitats. Coral reef bleaching is a striking example of how warmth and ocean acidity affect coastal and marine ecosystems. Along with other significant economic consequences, the Great Barrier Reef's bleaching in Queensland, Australia, may cause 10,000 jobs lost, AUD 1 billion in tourism income lost, and 1 million yearly visits lost. Damage from Florida's coral reefs might raise flood risk for around 7,300 homes, therefore costing USD 823 million year. Rising marine heat waves in the 1980s destroyed marine life, therefore compromising food supplies.

One of the concerns is environmental pollution. About 80% of wastewater is dumped into ecosystems untreated; marine garbage is expected to cost USD 21 billion worldwide by 2020. The blue economy runs economic risks including inflation and supply chain disruptions, demographic changes, political obstacles including corruption, and social problems resulting from differences between local and international fishing interests.

Survey data shows that coastal erosion (35%), floods (41%), and sea level rise (54%), are the main climatic hazards to the blue economy. The effects are major: sea level rise is expected to affect 800 million people in 570 coastal cities; floods cause yearly property damage of around USD 120 billion, by as much as 12% and by affecting up to 6% of the world GDP by 2025, which might impede local economy growth. Coastal erosion might cause a 5% drop-in coastal ecosystem services in Europe including shipping, tourism, and other coastal businesses.

Bennett et al. (2021) identified that pollution and waste-related environmental justice issues, environmental degradation and ecosystem service reduction are among the risks of ocean economy.

10.5.3 Economic and Social Challenges

The rhetoric and promises of the blue economy are seductive, however, a fairer and more just blue economy remains an uphill task.

The study of Bennett et al. (2021) dealt with the risks of ocean economy in their study on blue growth and blue justice. While more and more people see the seas as a potential new area for economic growth. However, significant threats to people and the environment may materialize as governments and businesses compete to profit from maritime resources. The authors conducted a literature review and highlight 10 potential societal injustices brought about by blue growth. Among the economic and social ones are 1) ocean grabbing, displacement, and dispossession; 2) harm to small-scale fishermen's livelihoods; 3) loss of access to marine resources necessary for food security and well-being; 4) unequal economic benefit distribution; 5) social and cultural effects; 6) marginalization of women; 7) violations of human and Indigenous rights; and 8) exclusion from governance.

10.5.4 Financial Constraints for Implementing Sustainable Innovations

A stable economy and long-term financial planning are prerequisites for a blue economy, which some nations find to be somewhat difficult given COVID-19. The blue economy is mostly implemented in underdeveloped nations that pay the cost; financial constraints play a major part in its application. Some underdeveloped nations have significant degrees of outside debt, so the emphasis will not be on changing their agricultural system to a bluer one. Some nations find the change more difficult because of their lack of ability and technology. Moreover, a nation needs a qualified workforce and thus field-based training.

The United Nations underlines that while promoting a blue economy, fairness cannot be overlooked. Land and resources frequently belong to communities, and the interests of those depending on the ocean are often neglected as major industries like coastal tourism are seen to generate more profit. The blue economy must therefore assist in reaching SDG 14 without concurrently undermining other 2030 Agenda targets.

10.5.4.1 Conceptual Framework Development

Given the critical implications of the global blue economy's rapid expansion, the economic sustainable use of oceanic resources is a critical challenge. Thus, blue growth is proposed as a resolution. The importance of ingenuity to achieve sustainable growth is examined and highlighted. The notion of sustainable innovation within the blue economy offers an extensive framework for comprehending the advancement and effects of sustainability efforts in marine and coastal domains. The framework

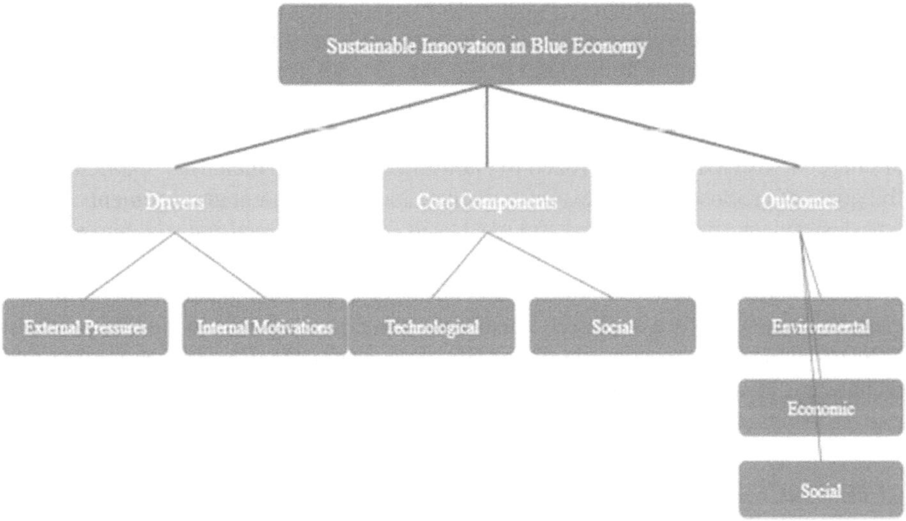

FIGURE 10.1 Conceptual framework: Sustainable innovation in blue economy.

Source: Author's completion.

is based on two principal variables: internal forces and external forces, including market demands, legal obligations, and environmental issues. Sustainable innovation pertains to innovations aimed at the dual goals of economic growth and environmental improvement. Sustainable innovation types include business model innovations, process innovations, product innovations, and managerial innovations. Figure 10.1 illustrates the conceptual framework for sustainable innovation in the blue economy which can help comprehend and facilitate sustainable innovations in any industry or economic sector's blue growth endeavors.

The framework depicts a conceptual structure for sustainable innovation in the blue economy, through three primary detachments—drivers, core components, and outcomes.

10.6 CONCLUSION

Thinking beyond scientific and technological innovations and the apparently never-ending work to learn more about the ocean are keystones on the road to a SBE.

In order to ensure and encourage sustainable investment in the blue economy and to provide investors with an understanding of their total longer-term prospective financial returns, the blue economy sustainability through innovation framework may assist and guide policy levers and actions.

The transition toward a SBE must be underpinned by a clear understanding of the opportunities presented by the ocean and how they should be evaluated, prioritized, and implemented. This will help people comprehend the blue economy idea and its significance in fostering economic development, achieving SDG 14, and enhancing people's quality of life by guaranteeing food security. Current economic modeling and scenario frameworks applied to ocean systems examine certain issues in great depth but ignore others entirely or only consider them at a very high level. A commitment to "blue justice" must be at the heart of the blue growth agenda (Bennett et al., 2021).

Technological, financial, legal, and economic concerns are largely interconnected, and collaboration between the many stakeholders is essential. Policymakers are urged to use a holistic approach to planning and management that: explicitly state the goals of proposed activities, investments, and management arrangements; prioritize desirable activities and investments based on clear, agreed criteria; and evaluate proposed activities against all relevant.

Additionally, creative technologies and business models along with creativity in governance frameworks are vital to compete and support development in the blue economy industry, which requires people to modify their professional profiles regarding new technology, IT tools, climate change concerns, renewable energy, and smart systems.

REFERENCES

African island nations at UN spotlight 'blue' economy, ocean's 'transformative power' as key to survival. (2016, September 26). *United Nations Sustainable Development*. https://www.un.org/sustainabledevelopment/blog/2016/09/african-island-nations-at-un-spotlight-blue-economy-oceans-transformative-power-as-key-to-survival/

Amon, D., Metaxas, A., Stentiford, G., Escovar-Fadul, X., Walker, T. R., Diana, Z., Karathanasi, F., Voyer, M., & Hemery, L. (2022). Blue economy for a sustainable future. *One Earth*, *5*(9), 960–963. https://doi.org/10.1016/j.oneear.2022.08.017

Bax, N., Novaglio, C., Maxwell, K., Meyers, K., Mccann, J., Jennings, S., Frusher, S., Fulton, E., Nursey-Bray, M., Fischer, M., Anderson, K., Layton, C., Emad, G. R., Alexander, K., Rousseau, Y., Lunn, Z., & Carter, C. (2022). Ocean resource use: Building the coastal blue economy. *Reviews in Fish Biology and Fisheries*, *32*, 1–19. https://doi.org/10.1007/s11160-021-09636-0

Bennett, N. J., Blythe, J., White, C. S., & Campero, C. (2021). Blue growth and blue justice: Ten risks and solutions for the ocean economy. *Marine Policy*, *125*, 104387. https://doi.org/10.1016/j.marpol.2020.104387

Buonomano, A., Giuzio, G. F., Maka, R., Palombo, A., & Russo, G. (2024). Empowering sea ports with renewable energy under the enabling framework of the energy communities. *Energy Conversion and Management*, *314*, 118693. https://doi.org/10.1016/j.enconman.2024.118693

Cisneros-Montemayor, A. M., Croft, F., Issifu, I., Swartz, W., & Voyer, M. (2022). A primer on the "blue economy": Promise, pitfalls, and pathways. *One Earth*, *5*(9), 982–986. https://doi.org/10.1016/j.oneear.2022.08.011

Falk-Andersson, J., Rognerud, I., De Frond, H., Leone, G., Karasik, R., Diana, Z., Dijkstra, H., Ammendolia, J., Eriksen, M., Utz, R., Walker, T. R., & Fürst, K. (2023). cleaning up without messing up: Maximizing the benefits of plastic clean-up technologies through new regulatory approaches. *Environmental Science & Technology*, *57*(36), 13304–13312. https://doi.org/10.1021/acs.est.3c01885

Hemer, M. A., Manasseh, R., McInnes, K. L., Penesis, I., & Pitman, T. (2018). Perspectives on a way forward for ocean renewable energy in Australia. *Renewable Energy*, *127*, 733–745. https://doi.org/10.1016/j.renene.2018.05.036

Lee, K. H., Noh, J., & Khim, J. S. (2020). The blue economy and the United Nations' sustainable development goals: Challenges and opportunities. *Environment International*, *137*, 105528.

Martínez-Vázquez, R. M., Milán-García, J., & de Pablo Valenciano, J. (2021). Challenges of the blue economy: Evidence and research trends. *Environmental Sciences Europe, 33*(1), 61. https://doi.org/10.1186/s12302-021-00502-1

Nikolopoulou, A. (2019). *Sustainable port development: Towards the Physical Internet concept.* https://www.academia.edu/68539295/Sustainable_port_development_towards_the_Physical_Internet_concept

OECD. (2024). *The Benefits and Costs of the Blue Economy at the Territorial Level.* OECD. https://doi.org/10.1787/134abd9f-en

Pace, L. A., Saritas, O., & Deidun, A. (2023). Exploring future research and innovation directions for a sustainable blue economy. *Marine Policy, 148*, 105433. https://doi.org/10.1016/j.marpol.2022.105433

Rotter, A., Giannakourou, A., Argente García, J. E., Quero, G. M., Auregan, C., Triantaphyllidis, G., Venetsanopoulou, A., De Carolis, R., Efstratiou, C., Aboal, M., Abad, M. E., Grigalionyte-Bembič, E., Kotzamanis, Y., Kovač, M., Ljubić Čmelar, M., Luna, G. M., Aguilera, C., Acién Fernández, F. G., Gómez Pinchetti, J. L., … Chiavetta, C. (2023). Identification of marine biotechnology value chains with high potential in the Northern Mediterranean region. *Marine Drugs, 21*(7), 416. https://doi.org/10.3390/md21070416

Stead, S. M. (2019). Using systems thinking and open innovation to strengthen aquaculture policy for the United Nations sustainable development goals. *Journal of Fish Biology, 94*(6), 837–844. https://doi.org/10.1111/jfb.13970

Techera, E. J., & Winter, G. (2021). The blue economy: Legal framework and tools for implementation. In *Marine Environmental Governance* (pp. 145–166). Springer.

United Nations. (2021). *The Sustainable Development Goals Report 2021.* United Nations. https://unstats.un.org/sdgs/report/2021/The-Sustainable-Development-Goals-Report-2021.pdf

United Nations Global Compact. (2020). *Ocean Stewardship 2030: Five Opportunities for Action to Ensure a Healthy and Productive Ocean.* UN Global Compact.

Visbeck, M. (2018). Ocean science research is key for a sustainable future. *Nature Communications, 9*(1), 690. https://doi.org/10.1038/s41467-018-03158-3

Webb, P. (2023). Introduction to oceanography. Licensed under a Creative Commons Attribution 4.0 International License. Retrieved from https://rwu.pressbooks.pub/webboceanography/chapter/1-1-overview-of-the-oceans/

Wenhai, L., Cusack, C., Baker, M., Tao, W., Mingbao, C., Paige, K., Xiaofan, Z., Levin, L., Escobar, E., Amon, D., Yue, Y., Reitz, A., Neves, A. A. S., O'Rourke, E., Mannarini, G., Pearlman, J., Tinker, J., Horsburgh, K. J., Lehodey, P., … Yufeng, Y. (2019). Successful blue economy examples with an emphasis on international perspectives. *Frontiers in Marine Science, 6*. https://www.frontiersin.org/articles/10.3389/fmars.2019.00261

Winther, J.-G., Dai, M., Rist, T., Hoel, A. H., Li, Y., Trice, A., Morrissey, K., Juinio-Meñez, M. A., Fernandes, L., Unger, S., Scarano, F. R., Halpin, P., & Whitehouse, S. (2020). Integrated ocean management for a sustainable ocean economy. *Nature Ecology & Evolution, 4*(11), 1451–1458. https://doi.org/10.1038/s41559-020-1259-6. https://doi.org/10.1038/s41559-020-1259-6

Yasser, M. M., Halim, Y. T., & Elmegaly, A. A. A. (2024). The blue economy effects on EUROMED tourism: Forecasting approach. *Future Business Journal, 10*(1), 100. https://doi.org/10.1186/s43093-024-00388-4

Yousef, M. (2024). Blue economy: A review of concepts, definitions, benefits, and risks. *American Academic & Scholarly Research Journal, 16*(4), 16–27. https://aasrc.org/aasrj/index.php/aasrj/article/view/2285

Part V

Human Capital and Social Dimensions

11 Role of Social Media in Creating Social Capital to Facilitate Human Capital and Sustainability in the Contemporary World

Shad Ahmad Khan and Ujjal Bhuyan

11.1 INTRODUCTION

Ushering in the era of Web 2.0 brought along a new revolution in communication between individuals. This two-way Internet was focused on the user experience, web participation, and content sharing. Initial and ongoing optimism regarding the Internet was founded on its ability to improve the lives of individuals with growing digital contact and more face-to-face interaction (Botella et al., 2012). Social media (SM), one of the building blocks of Web 2.0, is a user-generated innovation that allows content creation and sharing. Although mass usage of SM was encouraged by the rise of high-speed Internet, its roots go back to earlier platforms such as "Open Diary," which was created by Bruce and Susan Ableson in 1998 (Kaplan, 2011).

Kaplan and Haenlein (2010) define SM as a set of Internet-based technologies that enable the production and exchange of user-generated content and are built on the conceptual and technological pillars of Web 2.0. Boyd and Ellison (2018, p. 211) further expound on the explanation, defining SM as "web-based services that enable users to (1) create a public or semi-public profile in a bounded system, (2) define a list of other users with whom they share a connection, and (3) view and traverse their list of connections and those others have made within the system."

In the rapidly evolving digital era, SM sites have become integral to everyday communication, education, and socialization. Beyond their initial purpose of entertainment and social interaction, the sites now serve as dynamic spaces for civic action, knowledge exchange, and collective mobilization (Chen & Li, 2017; Utz & Muscanell, 2015). SM is a wide array of discussion boards and networking websites such as Facebook, Twitter, YouTube, blogs, and online forums (Pan et al., 2019), which distinguish themselves by enabling users to announce and post their social networks. SM is a crucial part of the formation of social capital (SC), thus advocating for human capital and sustainable development (Julsrud, 2023).

DOI: 10.1201/9781003611905-16

SM content and communication can influence people's perceptions, moods, and behaviors in the various capacities—in which they are citizens, consumers, or workers (Saleem et al., 2022; Wang et al., 2016; Thoumrungroje, 2014). In this context, individuals engage with SM networks based on personal preferences (Al-Bahrani et al., 2015). Usage of SM has turned into a valued social behavior, largely due to its potential to create interpersonal relationships (Sharma et al., 2023; Sharma et al., 2024). Yamakanith and Gurusamy (2014) research supports the argument that SM networking websites are capable of developing interpersonal relationships and, hence forming social connections. Such connections, referred to as "SC," play an inevitable role in social cohesion and resource sharing.

SC includes networks, norms, and trust that allow for coordination and cooperation for a common interest. SC gives access to information such as knowledge, information, and support, thereby strengthening interpersonal relationships (Adler & Kwon, 2002; Bino et al., 2022). Low SC societies also experience more social fragmentation, reduced levels of community involvement, and more mistrust. Dense social connections, however, enhance group mobilization and strengthen the feelings of community (Ellison et al., 2007).

In SM environments, SC is embodied as structural, relational, and cognitive features. These features influence the access to information of individuals and groups, their skills-building, and their sustainability behaviors. Human capital, framed as individuals' knowledge, skills, and abilities, is more and more magnified through such digitally facilitated social engagement. The theory of uses and gratification (UGT) assumes that consumption of media is driven by an individual's conscious action to meet specific needs and gratifications (Ruggiero, 2000). SM, in this case, is used as a method for fulfilling informational, social, and psychological needs such as seeking information, understanding community, feeling belongingness, and constructing identity (Hossain, 2019).

In light of the common challenges with respect to socio-economic disparities, environmental harm, and institutional limitations, the capacity to generate and leverage SC through the use of digital channels has become more imperative. An understanding of the significance of SM in shaping aspects of SC is therefore crucial to scholarly research and policy development in the contemporary knowledge economy.

This chapter investigates the function of SM in creating SC, i.e., its structural, relational, and cognitive dimensions and its contribution to the broader goals of human capital development and sustainable development. SM use is examined from social, cognitive, and hedonic standpoints. Empirically exploring SM use patterns and SC creation, utilizing partial least squares structural equation modeling (PLS-SEM), the study aims to reflect on the implications of the different patterns of SM usage in creating SC and how such implications could relate to sustainability in the world today.

11.2 SM USE AND SC: A MULTIDIMENSIONAL PERSPECTIVE

11.2.1 The Multifaceted Use of SM

SM such as Facebook, LinkedIn, Instagram, and Twitter are now essential communication tools that are transforming the manner in which individuals communicate with one another and with larger audiences (Blank & Lutz, 2017; Khan, 2023). These platforms enable diverse user requirements that can be summarized as "social, cognitive, and hedonic" usage dimensions (Cao & Yu, 2019; Katz et al., 1973; McQuail, 1994).

- **Social use (SSU)** is a description of using SM to build or maintain relationships, create new social ties, and link up with like-minded individuals (Jong et al., 2021; Quan-Haase & Young, 2010).
- **Cognitive usage (SCU)** is characterized by information consumption, sharing, and co-creation—ranging from knowledge search to the sharing of ideas and experiences via user-generated content (Rathore et al., 2016; Wang et al., 2022).
- **Hedonic use (SHU)**, however, deals with the use of SM for pleasure, recreation, and emotional gratification (Cao & Yu, 2019).

These three usage patterns are closely related to the UGT, which presumes that users deliberately choose specific media based on their expectations and needs (Ruggiero, 2000). UGT is a model of media consumption based on the user, implying that individuals actively select media platforms and content to fulfill particular psychological and social needs (Xu et al., 2025). Kaplan and Haenlein (2010) noted the collaborative and participatory character of SM, not just differentiating it from mass media but also amplifying the importance of SM in network building and collective resource mobilization.

11.2.2 SC: Concept, Dimensions, and Relevance

SC is traditionally used to describe the value created out of social relationships—has long been a theory that has captured the essence of community engagement and development (Leana & Pil, 2006; Paxton, 1999). SC, or social value of connection, has been a core indicator for examining community development and participation (Leana & Pil, 2006; Paxton, 1999). SC as conceptualized by Nahapiet and Ghoshal (1998) has three dimensions:

- **Structural social capital (SSC)** refers to the observable network ties and configuration of social relationships (McLeod et al., 2020).
- **Relational social capital** (RSC) is focused on trust, norms, and reciprocal obligations built through continuous social relationships; RSC is concerned with (McLeod et al., 2020).
- **Cognitive social capital (CSC)** is composed of common values, language, and knowledge that enable common interpretation and cooperation.

The relevance of SC is particularly significant in the age of the Internet. Ellison et al. (2007) presented how SM enables both bonding (strong relationships with acquaintances) and bridging (weak connections with diverse others) SC. Claridge (2018) also mentioned that the tangibility of structural SC makes it a valuable tool of analysis, especially when examining SM. Consequently, the current research employs the tripartite understanding of SC—structural, relational, and cognitive—since the aforementioned model is well-supported across contemporary literature (Mishchuk et al., 2022; Tajpour et al., 2022).

11.2.3 Connecting SM Use to SC

SM cognitive use (SCU) is the conscious and information-based utilization of SM platforms, for instance, searching for information, learning different perspectives, or participating in reflective conversation. SCU enhances the ability of individuals to process, share, and internalize information in useful terms, potentially building CSC by fostering shared meaning, values, and interpretations among SM users (Ode et al., 2025). In addition, as consumers interact cognitively with one another through smart conversation and knowledge exchange, it builds RSC since trust, reciprocity, and emotional ties based on respect and intellectual stimulation (H2). Furthermore, repeated and reflective communication among structured networks such as online forums, communities, or professional societies enhances SSC as they build and maintain relationships and interaction patterns in the SM environment (H3). Previous studies have demonstrated that knowledge-based SM behavior significantly promotes the development of well-developed and diverse social networks and meaningful relations, thereby verifying these hypotheses' suggestion.

H1: SCU has a positive influence on CSC.
H2: SCU has a positive influence on RSC.
H3: SCU has a positive influence on SSC.

Empirical studies suggest hedonic use is primarily recreational and is, in fact, an important source of social bonding and intellectual stimulation. In line with this, the following hypotheses are put forward:

H4: SHU has a positive influence on CSC.
H5: SHU has a positive influence on RSC.
H6: SHU has a positive influence on SSC.

The role of social media attachment (SMA) is referred as the emotional or habitual reliance on social platforms has also drawn interest. However, its relationship with SC appears more tenuous:

H7: SMA has a positive influence on CSC.
H8: SMA has a positive influence on RSC.
H9: SMA has a positive influence on SSC.

As SM has primarily social use (Hall, 2018), the same applies for the SM for SC as social usage is expected to influence SC due to its interpersonal focus, evidence also suggests its impact may be inconsistent across dimensions:

H10: SSU has a positive influence on CSC.

H11: SSU has a positive influence on RSC.
H12: SSU has a positive influence on SSC.

11.2.4 The Role of SM in Higher Education and Social Integration

SM usage is nearly universal among college students, serving as a source of both social integration and academic involvement (Greenhow & Burton, 2011). Websites allow students to maintain existing relationships, form new relationships, and feel more connected to peers and professors. Such feelings of belongingness and involvement have a direct influence on student outcomes—both academic and psychological (Brooman & Darwent, 2014; Dika & Singh, 2002; Saleem et al., 2023).

Support from Kamboj et al. (2017) attests to the contention that all three forms of SM use—social, hedonic, and cognitive—have the potential to significantly enhance the three facets of SC. These findings are aligned with earlier work by Pan et al. (2019), who highlighted how being online enhances social inclusion, thereby enabling human capital development.

While early skepticism linked SM to the downfall of traditional forms of engagement and a decline in SC (Nie, 2001), subsequent research has invalidated this view. This contention was advanced by Wellman (2001), who argued that online networks are more likely to enhance, rather than replace, face-to-face relationships. This "hybrid sociality" strategy emphasizes the possibility of SM functioning as a multiplier of social cohesion, knowledge sharing, and sustainable community building—even amidst moral panics about digital addiction (Cohen, 2011; Valenzuela et al., 2008).

11.3 METHODOLOGY

This study aims to investigate how SMA, SSU, and SCU influence SHU, SSC, RSC, and CSC through quantitative and cross sectional design. As the reliance of young people on the Internet continues to increase, the study focuses on the population to better understand the processes by which SM constructs SC and results in human and sustainable development.

Primary data were collected via an online formal questionnaire that yielded 93 valid responses. The demographic sample profile of the respondents illustrates a predominant ratio of the young as expected in the study aim. The gender distribution was 63.4% male (n = 59) and 36.6% female (n = 34). With regards to age groups, 69.9% were 18–22 years (n = 65), 25.8% were 22–26 years (n = 24), and 4.3% were 26–30 years (n = 4). The age group here represents a digitally engaged population, most likely to exhibit strong patterns of SM activity and SC formation.

To determine the adequacy of the sample size, a G*Power 3.1 analysis was conducted. Using an a priori power analysis for linear multiple regression with a medium effect size ($f^2 = 0.15$), $\alpha = 0.05$, power ($1-\beta$) = 0.80, and three predictors (SMA, SSU, SCU), the minimum sample size needed was calculated to be 77. The sample of 93 respondents obtained is above this, offering more than sufficient statistical power for efficient model testing. Moreover, using PLS-SEM with the SmartPLS software is justified as it is appropriate for small samples and for testing complex models with multiple constructs and mediating relationships.

All the constructs were measured by using pre-validated items from prior research translated into the setting of the study. Data was administered on a 7-point Likert scale, i.e., from 1 (strongly disagree) to 7 (strongly agree). The constructs examined included: SMA, SSU, SCU, SHU, SSC, RSC, and CSC.

The PLS-SEM analysis was conducted using a two-stage approach. In the first step, the measurement model's convergent and discriminant validity and internal consistency reliability (Cronbach's alpha and composite reliability) were evaluated. In the second step, the structural model was validated to test the hypothesized constructs' relationships through the inspection of path coefficients, t-values, p-values, R^2 values, and model fit assessment. This methodological approach provides for an end-to-end and statistically sound evaluation of the impact of consumption of SM on SC development in the context of human capital and sustainability.

11.4 DATA ANALYSIS

11.4.1 SM Usage

As shown in Table 11.1, gender trends in SM usage reflect divergent patterns. Males, at 36.6%, are heavy users, recording more than 3 hours of use per day, followed by 24.7% using it for 1–2 hours. Females, however, show a trend toward moderate usage, reporting 35.3% for 2–3 hours and a comparatively lower 27.9% under heavy usage. This indicates that males spend significantly more time on SM, while females use it in more balanced and structured ways.

Looking at age trends in Table 11.2, the 18–22 age group dominates the >3 hours usage category (36.9%), which could suggest that this age group has the most interaction with SM. The 22–26 year olds have a more even usage with 29.2% each in the 2–3 hour and 1–2 hour categories. Concurrently, the 26–30 age range shows a consistent pattern of declining use, 0% heavy use and 50% short, 31–60-min use. This shows SM use reducing with age, a departure from heavy to more controlled and structured usage.

TABLE 11.1
SM Usage by Gender (n=93)

Gender	>3 hours	2–3 hours	1–2 hours	31–60 min	10–30 min
Male (%)	36.60	22.60	24.70	11.80	4.30
Female (%)	27.90	35.30	20.60	14.70	1.50
Total (%)	**33.30**	**26.90**	**23.70**	**12.90**	**4.30**

TABLE 11.2
SM Usage by Age Group (n=93)

Age	>3 hours	2–3 hours	1–2 hours	31–60 min	10–30 min
18–22 (%)	36.90	26.20	21.50	10.80	4.60
22–26 (%)	25.00	29.20	29.20	16.70	0
26–30 (%)	0	25.00	25.00	50.00	0

TABLE 11.3
SM Usage by Education (n=93)

Education	>3 hours	2–3 hours	1–2 hours	31–60 min	10–30 min
Bachelor's Degree (%)	37.70	26.20	21.30	9.80	4.90
Master's Degree (%)	17.20	37.90	27.60	17.20	0
Other* (%)	33.30	22.20	22.20	22.20	0

Breaking down education level in Table 11.3, bachelor students lead heavy use (>3 hours: 37.7%), indicating unrestricted SM behavior. Master's students prefer 2–3 hours of use (37.9%), reflecting a more organized and goal-oriented pattern. Students in the "Other" category (Ph.D. and professional programs) reflect a well-distributed pattern across use categories with no bias toward heavy use (only 33.3% reported >3 hours). This reflects that higher-level education is associated with more organized and moderated use of SM.

At an overall level, SM usage information reveals that less-educated, more young men are predominant frequent users, while older and more educated individuals move toward moderation and time management. Women in all brackets consistently opt for medium-level usage, while affirming the more patterned gender-related trend.

11.4.2 Measurement Model

11.4.2.1 Internal Consistency (Reliability)

The internal consistency, i.e., reliability of each of the constructs was ascertained through Cronbach's Alpha and composite reliability (rho_c) and all of them were higher than the acceptable threshold of 0.70 (Gebremedhin et al., 2022; Khan et al., 2021), indicating acceptable to superb internal consistency in the measurement model presented in Table 11.4.

11.4.2.2 Convergent Validity

Convergent validity was tested using average variance extracted (AVE) and all constructs had AVE scores more than 0.50 (Gebremedhin et al., 2022; Khan et al., 2021),

TABLE 11.4
Internal Consistency Analysis

Construct	Cronbach's Alpha	Composite Reliability (rho_c)	Interpretation
SCU	0.71	0.821	Acceptable
CSC	0.889	0.924	Excellent
SHU	0.923	0.945	Excellent
SMA	0.775	0.848	Good
RSC	0.908	0.935	Excellent
SSU	0.874	0.908	Excellent
SST	0.863	0.902	Excellent

TABLE 11.5
Convergent Validity

Construct	AVE	Interpretation
SCU	0.535	Acceptable
CSC	0.752	Excellent
SHU	0.812	Excellent
SMA	0.529	Acceptable
RSC	0.784	Excellent
SSU	0.665	Good
SST	0.648	Good

which ensures that each construct accounts for greater than half of the variance in its indicators as presented in Table 11.5.

11.4.2.3 Factor Loadings

All the factors based on the reliability and validity values were accepted to proceed for further analysis, the details are presented in Table 11.6.

11.4.2.4 Discriminant Validity

Discriminant validity was also tested using data from both the Heterotrait-Monotrait (HTMT) ratio and Fornell-Larcker criterion. All the HTMTs were less than the threshold value of 0.90, and most were below 0.85, indicating good discriminant

TABLE 11.6
Factor Load

Construct	Item	Loadings	Interpretation
SCU	**C1**	0.658	Acceptable
	C2	0.736	Good
	C3	0.765	Good
	C4	0.762	Good
CSC	**CS1**	0.901	Excellent
	CS2	0.881	Excellent
	CS3	0.889	Excellent
	CS4	0.792	Strong
SHU	**H1**	0.902	Excellent
	H2	0.881	Excellent
	H3	0.926	Excellent
	H4	0.894	Excellent
SMA	**I2**	0.596	Borderline/problematic
	I3	0.763	Good
	I4	0.806	Good
	I5	0.767	Good
	I6	0.685	Acceptable

(Continued)

TABLE 11.6 *(Continued)*
Factor Load

Construct	Item	Loadings	Interpretation
RSC	**R1**	0.854	Strong
	R2	0.913	Excellent
	R3	0.893	Excellent
	R4	0.88	Excellent
SSU	**S1**	0.764	Good
	S2	0.84	Strong
	S3	0.848	Strong
	S4	0.886	Excellent
	S5	0.73	Acceptable
SST	**ST1**	0.832	Strong
	ST2	0.76	Good
	ST3	0.834	Strong
	ST4	0.846	Strong
	ST5	0.749	Acceptable

validity (Khan & Magd, 2021; Khan et al., 2021). Additionally, the square root of the AVE of every construct was higher than the corresponding inter-construct correlations, meeting the Fornell-Larcker criterion validity (Khan & Magd, 2021; Khan et al., 2021). The results confirm that the constructs are statistically distinct and represent different concepts in the model as presented in Table 11.7.

TABLE 11.7
Discriminant Validity

				Fornell Larker Criterion			
	CSC	RSC	SCU	SHU	SMA	SSC	SSU
CSC	0.867						
RSC	0.558	0.885					
SCU	0.454	0.639	0.732				
SHU	0.453	0.479	0.191	0.901			
SMA	0.505	0.432	0.443	0.42	0.727		
SSC	0.452	0.732	0.667	0.41	0.409	0.805	
SSU	0.489	0.504	0.483	0.55	0.534	0.561	0.816
				HTMT			
CSC							
RSC	0.622						
SCU	0.563	0.778					
SHU	0.496	0.509	0.264				
SMA	0.599	0.498	0.599	0.497			
SSC	0.519	0.827	0.838	0.457	0.488		
SSU	0.547	0.549	0.597	0.604	0.656	0.631	

11.4.2.5 Structural Model: Path Coefficients and Hypothesis Testing

The results show that both SCU and SHU of SM positively affect all three dimensions of SC, namely CSC, RSC, and SSC, since all the path coefficients are positive and statistically significant ($p < 0.05$). Among them, SCU has the highest influence on RSC and SSC ($\beta = 0.543$ and 0.542, respectively), followed by SHU with a moderate influence. On the other hand, SMA has no significant influence on any dimension of SC, with low t-values and high p-values, leading to the rejection of H7, H8, and H9. Similarly, SSU does not affect CSC or RSC but marginally positively affects SSC ($\beta = 0.198$, $p = 0.054$), resulting in the marginal support of H12. Collectively, the findings stress that strategic and socially oriented SM use results in SC formation, but attachment (including addiction) and self-focused use does not. The results are summarized in Table 11.8.

TABLE 11.8
Structural Model Analysis

Path	Coefficient (β)	T-Value	p-Value	2.50%	97.50%	Interpretation	Decision
SCU -> CSC	0.253	2.732	0.006	0.069	0.429	SCU has a positive influence on CSC.	H1 is accepted
SCU -> RSC	0.543	7.036	0	0.389	0.694	SCU has a positive influence on RSC.	H2 is accepted
SCU -> SSC	0.542	6.954	0	0.385	0.692	SCU has a positive influence on SSC.	H3 is accepted
SHU -> CSC	0.25	2.38	0.017	0.043	0.46	SHU has a positive influence on CSC.	H4 is accepted
SHU -> RSC	0.343	4.086	0	0.168	0.499	SHU has a positive influence on RSC.	H5 is accepted
SHU -> SSC	0.208	2.631	0.009	0.034	0.351	SHU has a positive influence on SSC.	H6 is accepted
SMA -> CSC	0.232	1.72	0.086	−0.028	0.499	SMA does not have a positive influence on CSC.	H7 is rejected
SMA -> RSC	0.027	0.258	0.796	−0.167	0.241	**H8**: SMA does not have a positive influence on RSC.	H8 is rejected
SMA -> SSC	−0.024	0.225	0.822	−0.221	0.195	**H9**: SMA does not have a positive influence on SSC.	H9 is rejected
SSU -> CSC	0.105	0.751	0.453	−0.157	0.389	**H10**: SSU does not have a positive influence on CSC.	H10 is rejected
SSU -> RSC	0.038	0.308	0.758	−0.196	0.283	**H11**: SSU does not have a positive influence on RSC.	H11 is rejected
SSU -> SSC	0.198	1.927	0.054	0.004	0.409	**H12**: SSU has a marginal positive influence on SSC.	H12 is marginally accepted

11.4.2.6 Coefficient of Determination and Model Fit

The structural model demonstrates satisfactory global fit and moderate to high explanatory power across the three dimensions of SC, i.e., SSC, RSC, CSC. The R^2 values for the endogenous variables indicate that the model explains 39.7% of the variance in CSC, 54.2% in RSC, and 54.8% in SSC, which represents a moderate to high explanatory power based on Cohen's (1988) guidelines. These results confirm that social use, hedonic use, cognitive use, and sustainability-oriented use of SM have a profound impact on the different dimensions of SC.

In terms of model fit, the SRMR (Standardized Root Mean Square Residual) measures how well the model's predicted correlations match the actual observed correlations. The SRMR indices (0.085 saturated and 0.091 estimated model) fall below the acceptable limit (<0.10), reflecting good approximation of the model to observed data. Additionally, distance-based metrics—d_ULS (3.543 sat., 4.115 est.) and d_G (1.713 sat., 1.787 est.)—are also in acceptable limits with negligible discrepancies between the saturated and estimated models, further showing good model fit.

Even though the normed fit index ranges between 0.662 and 0.672, a short of the optimum at 0.90, it's in a usually acceptable range for PLS-SEM, especially when using it for complicated models or non-normal data.

11.5 CONCLUSIONS

The study was able to examine the role of SM in building SC by determining the effects of various SM use patterns on structural, relational, and cognitive SC. The findings empirically confirm the significant role of how a person uses SM in building social relationships and values. In particular, the research demonstrates that SCU, i.e., knowledge sharing and information consumption, strongly and positively influences all three SC dimensions, particularly RSC and SSC. Similarly, SHU, driven by leisure and pleasure, also positively influences CSC, RSC, and SSC but with a weaker impact. These results are in line with hypotheses suggesting that both knowledge-based and recreational SM behaviors strengthen social network construction and intimate relations. SMA, having no meaningful positive effect on any SC dimension, was found. SSU, unexpectedly, did not have a significant influence on CSC or RSC, with only marginal positive influence on SSC. This highlights a key distinction: it is not merely being attached to or merely using SM socially, but the specific modes of cognitive and hedonic attachment that are key drivers of SC formation here. The study supports the hypothesis that SM can enhance, rather than replace, social bonding, with consequent "hybrid sociality." Lastly, the findings confirm that SM, used strategically and in socially aimed ways (especially cognitively and hedonically), is a valuable instrument for building all types of SC, which in turn supports human capital and sustainable development goals.

11.6 IMPLICATIONS

The study makes important theoretical and practical contributions to knowledge on the effects of SM use on SC formation. Theoretically, it adds to existing models in illustrating that specific patterns of SM use, i.e., SCU and SHU, have positive influences on SSC,RSC, and CSC, while attachment only or certain social uses do not, confirming again the relevance of uses and gratification theory and hybrid sociality. Practically, it welcomes individuals to utilize SM for learning and affective enrichment, suggests educators integrate SM into academic and social support networks, and advises policymakers to promote digital literacy that fosters productive online activity. For platform designers, the findings highlight the importance of features that facilitate meaningful engagement. Overall, the research emphasizes SM use with strategic purpose to formulate SC that is imperative for human capital development and sustainable social growth.

11.7 DIRECTION FOR FUTURE STUDIES

Based on the results and constraints of the research, several lines of future research are outlined. Longitudinal designs are suggested first to follow the dynamic evolution of the relationship between SM usage patterns and SC over time. Larger and more diverse samples for different age groups, occupations, and cultural backgrounds must be employed for the purposes of generalizability. Subsequent research also needs to explore mediators and moderators such as platform type, content nature, personality, and socio-cultural. The non-significant effect of SSU on SSC also needs more exploration into what aspects of SSU motivate network formation. The non-significant effect of SMA also necessitates further exploration into different attachment dimensions and their potential adverse effects. To better support the linkage of SM, human capital, and sustainability, the next wave of research needs to measure outcomes like skill building and community effects directly. Finally, qualitative or mixed-methods approaches might be able to produce richer insights about user experience and reveal how cognitive and hedonic interactions generate SC, thereby clarifying our understanding of hybrid sociality.

REFERENCES

Adler, P. S., & Kwon, S. (2002). Social capital: Prospects for a new concept. *Academy of Management Review*, *27*(1), 17–40. DOI:10.5465/AMR.2002.5922314

Al-Bahrani, A., Patel, D., & Sheridan, B. (2015). Engaging students using social media: The students' perspective. *International Review of Economics Education*, *19*, 36–50. https://doi.org/10.1016/j.iree.2015.06.001

Bino, E., Epoc, F., & Bystrov, I. (2022). Social capital and entrepreneurship. In M. Arafat, I. Saleem, J. Ali, A. Khan, & H. Balhareth (Eds.), *Driving Factors for Venture Creation and Success in Agricultural Entrepreneurship* (pp. 241–249). IGI Global Scientific Publishing.

Blank, G., & Lutz, C. (2017). Representativeness of social media in Great Britain: Investigating Facebook, LinkedIn, Twitter, Pinterest, Google+, and Instagram. *American Behavioral Scientist*, *61*(7), 741–756.

Botella, C., Riva, G., Gaggioli, A., Wiederhold, B. K., Alcaniz, M., & Banos, R. M. (2012). The present and future of positive technologies. *Cyberpsychology, Behavior, and Social Networking*, *15*(2), 78–84.

Boyd, D. M., & Ellison, N. B. (2018). Social network sites: Definition, history and scholarship. *Journal of Computer-Mediated Communication*, *13*, 210–230. https://doi.org/10.1111/j.1083-6101.2007.00393.x

Brooman, S., & Darwent, S. (2014). Measuring the beginning: A quantitative study of the transition to higher education. *Studies in Higher Education*, *39(9)*, 1523–1541. https://doi.org/10.1080/03075079.2013.801428.

Cao, X., & Yu, L. (2019). Exploring the influence of excessive social media use at work: A three-dimension usage perspective. *International Journal of Information Management*, *46*, 83–92. https://doi.org/10.1016/j.ijinfomgt.2018.11.019

Chen, H. T., & Li, X. (2017). The contribution of mobile social media to social capital and psychological well-being: Examining the role of communicative use, friending and self-disclosure. *Computers in Human Behavior*, *75*, 958–965.

Claridge, T. (2018). Dimensions of social capital-structural, cognitive, and relational. *Social Capital Research*, *1*, 1–4.

Cohen, S. (2011). *Folk Devils and Moral Panics*. Routledge.

Dika, S. L., & Singh, K. (2002). *Applications of social capital in educational literature: A critical synthesis. Review of Educational Research*, *72*(1), 31–60. https://doi.org/10.3102/00346543072001031

Ellison, N. B., Steinfield, C., & Lampe, C. (2007). The benefits of Facebook "friends:" Social capital and college students' use of online social network sites. *Journal of Computer-Mediated Communication*, *12*(4), 1143–1168.

Gebremedhin, M., Gebrewahd, E., & Stafford, L. K. (2022). Validity and reliability study of clinician attitude towards rural health extension program in Ethiopia: Exploratory and confirmatory factor analysis. *BMC Health Services Research*, *22*(1), 1088.

Greenhow, C., & Burton, L. (2011). Help from my "friends": Social capital in the social network sites of low-income students. *Journal of Educational Computing Research*, *45*(2), 223–245.

Hall, J. A. (2018). When is social media use social interaction? Defining mediated social interaction. *New media & Society*, *20*(1), 162–179.

Hossain, M. A. (2019). Effects of uses and gratifications on social media use: The Facebook case with multiple mediator analysis. *PSU Research Review*, *3*(1), 16–28. https://doi.org/10.1108/PRR-07-2018-0023

Jong, D., Chen, S.-C., Ruangkanjanases, A., & Chang, Y.-H. (2021). The impact of social media usage on work efficiency: The perspectives of media synchronicity and gratifications. *Frontiers in Psychology*, *12*, 693183. https://doi.org/10.3389/fpsyg.2021.693183

Julsrud, T. E. (2023). Sustainable sharing in local communities: Exploring the role of social capital. *Local Environment*, *28*(6), 811–827.

Kamboj, S., Kumar, V., & Rahman, Z. (2017). Social media usage and firm performance: The mediating role of social capital. *Social Network Analysis and Mining*, *7*(1), 1–14. https://doi.org/10.1007/s13278-017-0468-8

Kaplan, A. M. (2011). Twitter ou le pouvoir de 140 caractères. *L'Expansion Management Review*, *140*(1), 104–113.

Kaplan, A. M., & Haenlein, M. (2010). Users of the world, unite! The challenges and opportunities of social media. *Business Horizons*, *53*(1), 59–68. https://doi.org/10.1016/j.bushor.2009.09.003

Katz, E., Haas, H., & Gurevitch, M. (1973). On the use of the mass media for important things. *American Sociological Review*, 164–181.

Khan, S. A. (2023). E-marketing, e-commerce, e-business, and internet of things: An overview of terms in the context of small and medium enterprises (SMEs). In A. Naim & V. Devi (Eds.), *Global Applications of the Internet of Things in Digital Marketing* (pp. 332–348). IGI Global.

Khan, S. A., Epoc, F., Gangwar, V., Ligori, T. A. A., & Ansari, Z. A. (2021). Will online banking sustain in Bhutan post COVID-19? A quantitative analysis of the customer e-satisfaction and e-loyalty in the Kingdom of Bhutan. *Transnational Marketing Journal*, *9*(3), 607–624.

Khan, S. A., & Magd, H. (2021). Empirical examination of MS teams in conducting webinar: Evidence from international online program conducted in Oman. *Journal of Content, Community and Communication*, *14*(7), 159–175.

Leana, C. R., & Pil, F. K. (2006). Social capital and organizational performance: Evidence from urban public schools. *Organization Science*, *17*(3), 353–366.

McLeod, J., Adams, A., & Sang, K. (2020). Ethical strategists in Scottish football: The role of social capital in stakeholder engagement. *International Journal of Business Governance and Ethics*, *14*(3), 298–316.

McQuail, D. (1994). *Mass Communication Theory*. SAGE.

Mishchuk, H., Štofková, J., Krol, V., Joshi, O., & Vasa, L. (2022). Social capital factors fostering the sustainable competitiveness of enterprises. *Sustainability*, *14*(19), 11905.

Nahapiet, J., & Ghoshal, S. (1998). Social capital, intellectual capital, and the organizational advantage. *Academy of Management Review*, *23*(2), 242–266. https://eli.johogo.com/Class/p18.pdf

Nie, N. H. (2001). Sociability, interpersonal relations, and the internet: Reconciling conflicting findings. *American Behavioral Scientist*, *45*(3), 420–435.

Ode, E., Awolowo, I. F., Nana, R., & Olawoyin, F. S. (2025). Social capital and artificial intelligence readiness: The mediating role of cyber resilience and value construction of SMEs in resource-constrained environments. *Information Systems Frontiers*, 1–21.

Pan, Y., Coleman, L., Manago, S., & Goodof, D. (2019). Effects of social media usage on social integration of university students. *International Journal of Technology in Teaching and Learning*, *15*(1), 1–17.

Paxton, P. (1999). Is social capital declining in the United States? A multiple indicator assessment. *American Journal of Sociology*, *105*(1), 88–127.

Quan-Haase, A., & Young, A. L. (2010). Uses and gratifications of social media: A comparison of Facebook and instant messaging. *Bulletin of Science, Technology & Society*, *30*(5), 350–361.

Rathore, A. K., Ilavarasan, P. V., & Dwivedi, Y. K. (2016). Social media content and product co-creation: An emerging paradigm. *Journal of Enterprise Information Management*, *29*(1), 7–18.

Ruggiero, T. E. (2000). Uses and gratifications theory in the 21st century. *Mass Communication and Society*, *3*(1), 3–37, https://doi.org/10.1207/S15327825MCS0301_02

Saleem, M., Khan, S. A., Al Shamsi, I. R., & Magd, H. (2023). Digital marketing through social media best practices: A case study of HEIs in the GCC region. In *Global Applications of the Internet of Things in Digital Marketing* (pp. 17–30). IGI Global.

Saleem, M., Khan, S. A., & Magd, H. (2022). Content marketing framework for building brand image: A case study of Sohar International School, Oman. In *Building a Brand Image through Electronic Customer Relationship Management* (pp. 64–83). IGI Global.

Sharma, N., Khatri, B., Khan, S. A., & Shamsi, M. S. (2023). Extending the UTAUT model to examine the influence of social media on tourists' destination selection. *Indian Journal of Marketing*, *53*(4), 47–64.

Sharma, N., Khatri, B., & Khan, S. A. (2024). Exploring the past and identifying the future research agenda for social media and tourist behavior: A bibliometric analysis. In C. Ramos, T. Costa, F. Severino, & M. Calisto (Eds.), *Social Media Strategies for Tourism Interactivity* (pp. 261–284). IGI Global.

Tajpour, M., Salamzadeh, A., Salamzadeh, Y., & Braga, V. (2022). Investigating social capital, trust and commitment in family business: Case of media firms. *Journal of Family Business Management*, *12*(4), 938–958.

Thoumrungroje, A. (2014). The influence of social media intensity and EWOM on conspicuous consumption. *Procedia-Social and Behavioral Sciences*, *148*, 7–15. https://doi.org/10.1016/j.sbspro.2014.07.009

Utz, S., & Muscanell, N. (2015). Social media and social capital: Introduction to the special issue. *Societies*, *5*(2), 420–424.

Valenzuela, S., Park, N., & Kee, K. F. (2008, April). Lessons from Facebook: The effect of social network sites on college students' social capital. In *9th International Symposium on Online Journalism* (pp. 4–5).

Wang, Y., Min, Q., & Han, S. (2016). Understanding the effects of trust and risk on individual behavior toward social media platforms: A meta-analysis of the empirical evidence. *Computers in Human Behavior*, *56*, 34–44. https://doi.org/10.1016/j.chb.2015.11.011

Wang, Y., Zhang, M., Luo, N., & Guo, L. (2022). Understanding how participating behaviours influenced by individual motives affect continued generating behaviours in product-experience-shared communities. *Behaviour & Information Technology*, *41*(14), 3044–3064.

Wellman, B. (2001). Computer networks as social networks. *Science*, *293*(5537), 2031–2034.

Xu, X. Y., Tayyab, S. M. U., Jia, Q., & Huang, A. H. (2025). A multi-model approach for the extension of the use and gratification theory in video game streaming. *Information Technology & People*, *38*(1), 137–179.

Yamakanith, P., & Gurusamy, P. (2014). A study on the influence of social networking sites on the interpersonal relationships of college students in Chennai and Coimbatore cities of India. *International Journal of World Research*, *1*(11), 48–56.

12 Readiness of Students and Employees in Logistics to Implement Digitalization, Sustainability, and Industry 4.0 in Practice

Simona Šinko, Beata Mrugalska, and Brigita Gajšek

12.1 INTRODUCTION

Logistics refers to the planning and execution of activities related to transport, warehousing, inventory, packaging, and the flow of information (Islam, 2018; Khan et al., 2019). Empirical studies have demonstrated a strong correlation between logistics performance and national economic growth and development, underscoring its significance as a critical enabler of competitiveness (Wong & Tang, 2018; Zhang et al., 2021). However, in recent decades, the logistics sector has undergone substantial transformation. The main drivers of change are Industry 4.0 technologies and a growing need to adapt to new market and operational requirements. Advances in digital technologies have facilitated the development of "Smart Logistics" or "Logistics 4.0," characterized by real-time data integration, automation, and cyber-physical systems (Gammelgaard, 2019; Winkelhaus & Grosse, 2020). Additionally, the growing influence of sustainability-oriented policies continues to reshape the logistics landscape by promoting environmentally responsible practices and regulatory compliance (Pacher et al., 2022).

In this evolving context, the availability of skilled workforce has emerged as a key determinant of success (Thai, 2012). Human resources constitute a fundamental pillar of logistics operations, and a strategic emphasis on developing a skilled workforce is necessary for adapting to ongoing structural and technological transformations (Cattaneo et al., 2017). Knowledge and knowledge management play a critical role in achieving operational excellence, enabling increased speed, quality, cost efficiency, and flexibility (Neumann & Tomé, 2006). However, the increasing integration of automation, digitalization, and sustainability-oriented practices introduces new complexities that logistics professionals and future entrants to the field must be adequately equipped to navigate.

 DOI: 10.1201/9781003611905-17

Compared to traditional logistics business models, contemporary logistics enterprises require highly qualified personnel, including managers and operational staff, who possess multidisciplinary expertise across domains such as logistics, finance, international trade, information technology, and communication (He et al., 2023). This need is increasingly critical as the required skill sets are evolving rapidly and must be aligned with the lifelong learning competencies outlined by the European Commission (2007). Despite these evolving demands, finding appropriately skilled logistics professionals remains a persistent challenge (von Randow, 2008). This challenge is further compounded by the fact that logistics students and professionals often focus primarily on the engineering aspects of logistics. As a result, sustainability-related issues are sometimes overlooked despite their growing importance in modern logistics practices (Cvahte Ojsteršek et al., 2023).

To address this gap and foster workforce readiness, it is important to establish the continuous development of capabilities related to the management of complex technological systems and the application of social intelligence in decision-making processes (Gupta et al., 2022). The logistics workforce is moving away from predominantly manual labour towards supervisory functions that require oversight of automated, data-driven systems (Parham, 2018). This transformation demands not only advanced technical expertise but also behavioural adaptability and organizational preparedness to effectively operate within an increasingly digitalized and dynamic work environment.

Despite a substantial body of literature on logistics-related competencies, the majority of studies continue to emphasize general business, management, and traditional logistics skills (Thai et al., 2011). However, these studies often fail to address the specific competencies required to meet the emerging demands of Industry 4.0, digitalization, and sustainability. For example, Kotzab et al. (2018) identified more than 280 relevant skills and competencies for logistics and supply chain managers. Nevertheless, analyses of job advertisement indicate considerable variation in competencies requirements depending on organizational level and function. As Gupta et al. (2022) noted, readiness for Industry 4.0 in logistics is shaped by three interrelated dimensions: organizational, behavioural, and technological. Their study of Indian logistics companies, using the fuzzy analytical hierarchical process, revealed that the existing workforce remains inadequately prepared to adopt and implement digital technologies comprehensively across all logistics operations. These findings point to the urgent need to reassess workforce capabilities and to make strategic investments in the development of competencies that align with the evolving technological and sustainability imperatives of modern logistics systems.

This study seeks to addresses the identified gaps by examining the readiness of both current logistics professionals (employees) and future professionals (students) for the digital and sustainable transformation of the sector. Specifically, it compares the self-assessed preparedness of students and employees with external evaluations provided by leading industry experts and representatives of national logistics associations (NLAs). The primary objective is to assess the extent to which these individuals are equipped to perform specific activities that have been identified in prior research as essential, based on a framework of key competencies previously published in the literature (Cvahte Ojsteršek et al., 2023).

The following sections of this chapter present the research methodology, results, and discussion. The methodology section describes the survey design and statistical procedures employed to evaluate perceived capabilities and readiness among logistic students and employees to implement key activities associated with Industry 4.0, digitalization, and sustainability. The results section provides a comprehensive analysis of both self-assessed readiness and external evaluations conducted by industry experts. The discussion section interprets these findings within broader context of workforce development, offering practical insights and strategic recommendations for higher educational institutions and industry stakeholders. The chapter concludes by outlining key theoretical and practical implications, along with proposed directions for future research in the domains of logistics education and workforce preparedness.

12.2 METHODOLOGY

This study investigates the readiness of both current logistics professionals and logistics students (future entrants to the profession) to implement tasks associated with Industry 4.0, digitalization, and sustainability in practical logistics settings. The research specifically compares respondent's self-assessed capabilities with external evaluations provided by representatives of leading companies recognized for their engagement with Industry 4.0 and sustainability initiatives, as well by NLAs.

This study is guided by the following research questions:

- Which core competencies related to digitalization, Industry 4.0, and sustainability are currently demonstrated by students and employees in the logistics sector?
- Which competences require further development to enhance logistics personnel readiness for application in practical logistics settings?
- What statistically significant differences can be observed between the self-assessed readiness of students and logistics professionals and the external assessments provided by representatives of industry and logistics associations concerning their ability to perform previously identified tasks?

The foundation of the survey design was established through a prior Delphi study conducted by Cvahte Ojsteršek et al. (2023) in which experts from six countries identified essential learning outcomes for logistics professionals in the context of Industry 4.0, digitalization, and sustainability. These learning outcomes were operationalized into 49 specific logistics-related activities, categorized as follows: 16 activities related to Industry 4.0, 8 to digitalization, and 25 to sustainability. This structured activity framework formed the basic of the online questionnaire used in the present study.

The survey consisted of two sections. The first section, which is analysed in this study, evaluated respondents' perceived ability to implement listed activities using a 4-point ordinal scale (1 = definitely not capable, 2 = probably not capable, 3 = probably capable, and 4 = definitely capable). Surveyed logistics employees and students were asked: "How capable do you feel about implementing following activities in practice?" Meanwhile, representatives from companies and NLAs responded to the question: "How would you evaluate the capability of an average national logistics employee to implement following activities in practice?" The second section of the survey aimed to identify the most suitable learning methods for acquiring the necessary knowledge.

The survey was conducted using the 1KA online Slovenian platform between March 8 and August 30, 2023. Participation was voluntarily and without incentives. The survey link for logistics students and employees was distributed to individuals from various faculties offering logistics programmes, as well from companies operating in the logistics sector. In contrast, participants representing exemplary cases of Industry 4.0 and sustainability implementation were carefully selected based on their companies' recognized achievements in these areas. The survey link was specifically sent to middle or senior management within these companies to ensure that the responses reflected the expertise of individuals directly involved in advanced logistics practices. Additionally, the heads of NLAs in the target countries were invited to participate, offering strategic, industry-wide insights. All participation was anonymous, fully GDPR-compliant, and intended solely for research purposes.

A total of 1,601 individuals accessed the introductory part of the survey, with 48% (763 participants) proceeding to the first page. Ultimately, 623 valid responses were collected, yielding a response rate of 38.9%. The majority of respondents (82.74%) were students, future logistics professionals, followed by individuals already employed in the logistics sector; participants represented a broad geographical range. The majority were from Poland, Portugal, Slovenia, and Turkey. Of the respondents, 14.68% came from non-partnering countries, including Austria, Bulgaria, Bosnia and Herzegovina, Croatia, Czech Republic, Germany, Hungary, Italy, the Netherlands, Romania, Russia, Spain, the United Kingdom, and Ukraine. A further 4.1% were from non-European countries such as India, Brazil, Iran, Jordan, Mexico, and the United States. A detailed distribution of respondents is presented in Table 12.1.

TABLE 12.1
Distribution of Respondents by Profile and Country

		Employee	NLA[1] Representative	Industry 4.0 Best Practice Representative	Sustainability Best Practice Representative	Student	Total
Country	Austria	5	0	1	0	7	13
	Italy	3	1	1	1	5	11
	Poland	29	1	3	4	148	185
	Portugal	11	2	4	1	86	104
	Slovenia	9	1	1	1	73	85
	Turkey	9	1	4	1	140	155
	Other	9	0	1	3	54	67
Total		75	6	15	11	513	620

[1] NLA – National logistic association

Statistical analysis was performed using IBM SPSS Statistics (29.0). Due to the ordinal nature of the data and its non-normal distribution, non-parametric tests were used. The Kruskal–Wallis H test was used to detect differences among respondent groups, followed by Mann–Whitney U tests for pairwise comparisons. These analyses aimed to examine variations in perceived readiness across respondent profiles and to identify capability gaps in executing key logistics activities.

12.3 RESULTS

This chapter presents the study's results, emphasizing the readiness of logistics students and employees to carry out logistics activities. It also highlights statistically significant differences between the self-assessment of students and employees and the capability evaluations conducted by representatives of technologically leading companies and NLA.

12.3.1 Readiness of Logistics Students and Employees to Apply Industry 4.0 Activities in Practice

Figure 12.1 illustrates respondents' self-assessment or their employees' capability to implement Industry 4.0-related logistics activities. Only valid responses, excluding "unable to respond," are included.

The highest confidence in capability was reported for the activity "Understand the design and deployment of artificial intelligence," with 64.79% of respondents indicating that they or their employees are either definitely or probably capable of implementation. Similarly, strong confidence was expressed in the ability to "Optimize logistics processes/routes/supply chains and use at least one optimization tool" (63.97% definitely or probably capable) and to "Explain Industry 4.0 – including its definition, implementation in logistics, forecasting basis, challenges, and pitfalls" (57.04% definitely or probably capable). Additionally, more than 55% of respondents rated themselves or their employees as capable of "using drones to support Industry 4.0/green/digitalization goals" (56.97%), "employing digital twins" (56.13%), and "conducting logistics process simulations" (55.88%).

The activity "Introduce autonomous warehouse equipment" had the highest proportion of respondents indicating that they were definitely not capable, at 21.91%. More than half (53.29%) reported being definitely or probably not capable of "Assessing the Industry 4.0 maturity level of their logistics department or company." Similarly, 51.79% expressed low confidence in "deciding on robot/cobot implementation." Significant uncertainty was also evident for activities such as "Plan warehouse automation" and "Decide on using machine learning," with over a quarter of respondents selecting probably not capable.

To identify statistically significant differences between respondent groups, statistical tests were conducted. Table 12.2 presents activities for which significant differences between respondent groups were observed.

Statistically significant differences emerged for multiple activities. For example, NLA representatives (mean rank = 249.30) and representatives of Industry 4.0 leaders (225.55) rated logistics employees' capability to "Introduce Logistics 4.0" significantly

FIGURE 12.1 Readiness to implement Industry 4.0-related logistics activities.

higher than students perceived their abilities (164.72 and 166.75, respectively). A similar pattern was observed for "Adopt the Industrial Internet of Things," where NLA representatives (237.10) again rated capabilities higher than students (158.26).

Employees (mean rank = 220.51) and representatives of sustainability leaders (230.75) expressed significantly greater confidence in "Implementing optimization tools to achieve sustainability goals" compared to students (187.62 and 163.88,

TABLE 12.2
Perceived Capability Differences by Group (Mann–Whitney U Test)

Activity	Mean Ranks by Group	Mann–Whitney U Test
Introduce Logistics 4.0 into the company's logistics	NLA representative (249.30) – Student (164.72)	398.500*
	I4.0 best practice representative (225.55) – Student (166.75)	1,059.500*
Adopt the Industrial Internet of Things (IIoT) in the operation of the logistics department/company	NLA representative (237.10) – Student (158.26)	394.500*
Implement optimization of logistics processes/routes/supply chain to achieve sustainability goals and use at least one optimization tool	Employee (220.51) – Student (187.62)	8,410.00*
	Sustainability best practice representative (230.75) – Student (163.88)	766.00*
Plan warehouse automation on a scale from partial automation to a dark warehouse based on economic feasibility	Sustainability best practice representative (224.19) – Student (159.90)	754.50*
Decide on the possibility and appropriateness of using cloud computing technologies to approach I4.0/green/digitalization goals of the logistics department/company	Employee (213.38) – Student (183.78)	7,987.00*
	NLA representative (245.33) – Student (159.91)	445.00*
Carry out an economic analysis of interventions to achieve sustainability goals and I4.0 goals	NLA representative (258.00) – Student (161.21)	387.00**

** The significance level is 0.01
* The significance level is 0.05

respectively). Similarly, for "Planning warehouse automation," sustainability best practice representatives (224.19) evaluated higher employees' ability than students (159.90).

For "Using cloud computing," both employees (mean rank = 213.38) and NLA representatives (245.33) reported greater confidence than students (183.78 and 159.91, respectively). The most notable difference was observed in "Carrying out an economic analysis of interventions," where NLA representatives (258.00) rated significantly higher than students (161.21), showing the strongest level of statistical significance ($p < 0.01$).

12.3.2 Readiness of Logistics Students and Employees to Apply Digitalization Activities in Practice

Figure 12.2 presents the results on the perceived capability to implement digitalization-related activities in practice, including all respondent group. For every listed activity, at least 50% of respondents indicated that they or logistics employees are either definitely or probably capable of practical implementing.

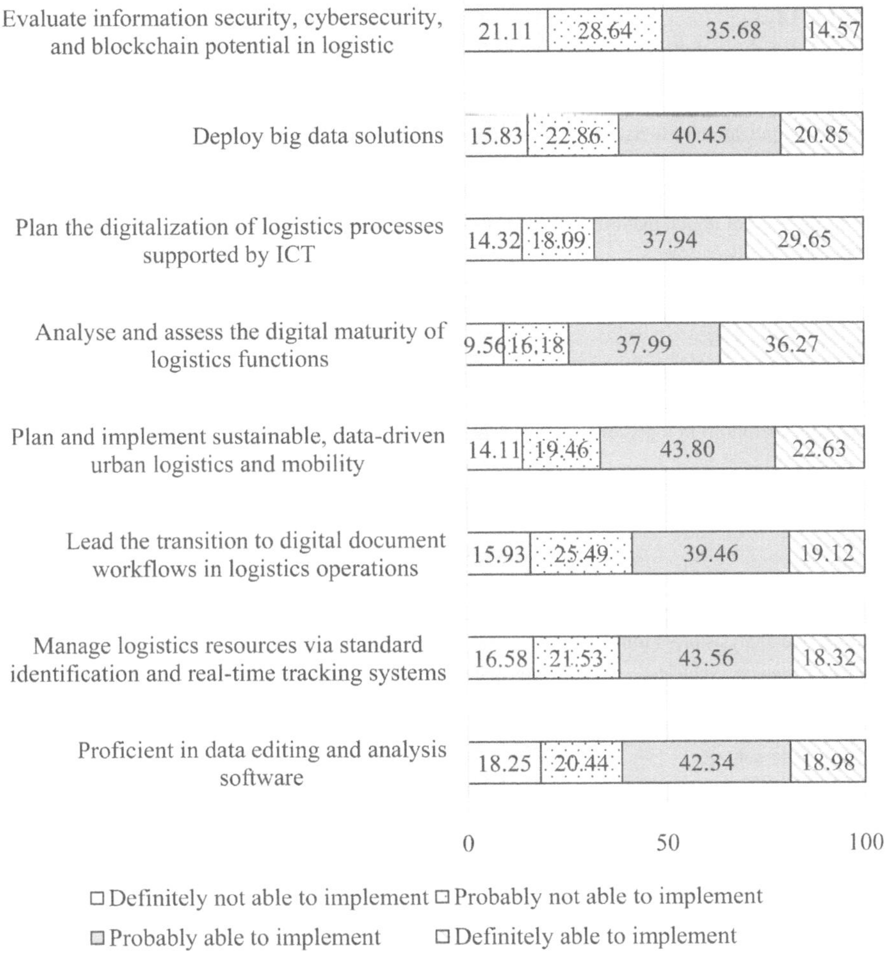

FIGURE 12.2 Readiness to implement digitalization-related logistics activities.

The highest confidence level was reported for the activity "Assess the level of digital maturity of the logistics department/company," with 74.26% of respondents indicating that they or logistics employees are capable of implementation. Additionally, 61.3% expressed confidence (probably or definitely capable) in "implementing big data analytics." In contrast, responses were more varied for activities related to the digitalization of logistics processes and sustainability-focused digital solutions. While approximately one-third of respondents indicated that they were definitely capable of implementing these activities, a considerable proportion expressed uncertainty.

Conversely, the lowest perceived capability was reported for activities related to assessing information and cybersecurity and using blockchain technologies. Specifically, 49.75% of respondents indicated that they or logistics employees were definitely or probably not capable of implementing these activities.

TABLE 12.3
Perceived Capability Differences by Group (Mann–Whitney U Test)

Activity	Mean Ranks by Group	Mann–Whitney U Test
Plan the digitalization of logistics processes with the support of ICT	Employee (219.29) – Student (189.08)	8,612.50*
	NLA representative (237.50) – Student (164.17)	540.00*
Assess the level of digital maturity of the logistics department/company	Employee (218.58) – Student (185.34)	7,813.00*
Introduce the collection and analysis of big data	Employed (224.18) – Student (187.24)	7,879.50*
Plan and lead digitalization of document flows and transition to paperless operations in logistics	Employed (32.46) – NLA representative (49.67)	277.00*
	Employed (33.33) – Sustainability best practice representative (47.75)	346.00*
	NLA representative (261.83) – Student (164.75)	406.00*
	Sustainability best practice representative (245.13) – Student (165.60)	683.00*
Advanced use of at least one data editing and analysis programme	Employee (226.26) – Student (186.98)	7,654.50**
Manage resources (vehicles, people, and logistics units) in logistics based on identification system and tracking	Employee (215.31) – Student (182.69)	7,911.00*
Plan and implement sustainable and information-supported urban logistics and people's mobility	Sustainability best practice representative (229.81) – Student (160.80)	725.50*
Assess the relevance of information and cybersecurity and the potential for using blockchain	NLA representative (248.90) – Student (160.12)	355.50*

** The significance level is 0.01

* The significance level is 0.05

A deeper analysis using the Mann–Whitney U test revealed statistically significant differences between students and employees' groups across eight activities (Table 12.3). Only statistically significant results are reported, with mean ranks provided for each comparison.

For the activity "Plan the digitalization of logistics processes with the support of information and communication technology," statistically significant differences were observed between students (mean rank = 189.08 and 164.17) and both logistics employees (mean rank = 219.29) and NLA representatives (237.50), with the latter groups reporting higher capability levels than students.

For the activity "Assess the level of digital maturity," logistics employees (218.58) rated their capability higher than students (185.34). A similar pattern was observed for "Introduce the collection and analysis of big data," with employees (224.18) again reporting higher capability than students (187.24). In the case of "Plan and lead digitalization of document flows and transition to paperless operations in logistics," several significant differences emerged. Employees had lower mean ranks (32.46 and 33.33) compared to NLA representatives (49.67) and sustainability best practice representatives (47.75). Moreover, both of the groups reported higher capability (164.75 and 165.60) than students (261.83 and 245.13, respectively).

For the activity "Advanced use of at least one data editing and analysis programme," logistics employees (mean rank = 226.26) again rated their capabilities significantly higher than students (186.98), with the result statistically significant at the $p < 0.01$. A similar significant difference was observed for "Manage resources in logistics based on identification and tracking technologies," where employees (215.31) rated themselves as more capable than students (182.69) did. Statistically significant difference also emerged for the activity "Plan and implement sustainable and information-supported urban logistics and people's mobility," with sustainability best practice representatives (229.81) evaluating their capabilities higher than students (160.80). Finally, for "Assess the relevance of information and cybersecurity and the potential for using blockchain," NLA representatives (248.90) indicated significantly higher capability for logistics employees compared to students (160.12).

12.3.3 Readiness of Logistics Students and Employees to Apply Sustainability Activities in Practice

Figure 12.3 presents the results related to the practical capability of implementing sustainability-related activities across all respondent groups. The highest proportion of respondents (30.86%) indicated that they or logistics employees are definitely capable of implementing was recorded for "Implement circular economy concepts in the logistics department/company." Additionally, 44.94% reported that they are probably capable of doing so.

For several activities, over 60% of respondents indicated that they or logistics employees are either definitely or probably capable of implementing them. These activities include "Implement circular economy concepts in the logistics department/company" (75.80%), "Make production logistics green" (70.00%), "Choose sustainable tangible fixed assets in logistics" (68.47%), "Assess the level of green maturity of the logistics department/company" (67.68%), "Prepare the logistics department/company to implement the reporting process in accordance with the GRI G4 guidelines" (66.75%), "Reduce energy consumption, environmental waste, and protect ecosystems in logistics" (66.92%), "Calculate the carbon footprint for logistics activity" (66.83%), "Manage social sustainability of the logistics department/company" (66.58%), "Check compliance of logistics activities with legal environmental regulations" (63.82%), and "Guide logistics behaviour in accordance with regulatory and policy frameworks for sustainable logistics" (63.78%).

For the activity "Manage economic sustainability of the logistics department/company," the highest percentage of respondents reported that they or logistics employees are definitely not capable of implementing it (16.58%). When considering

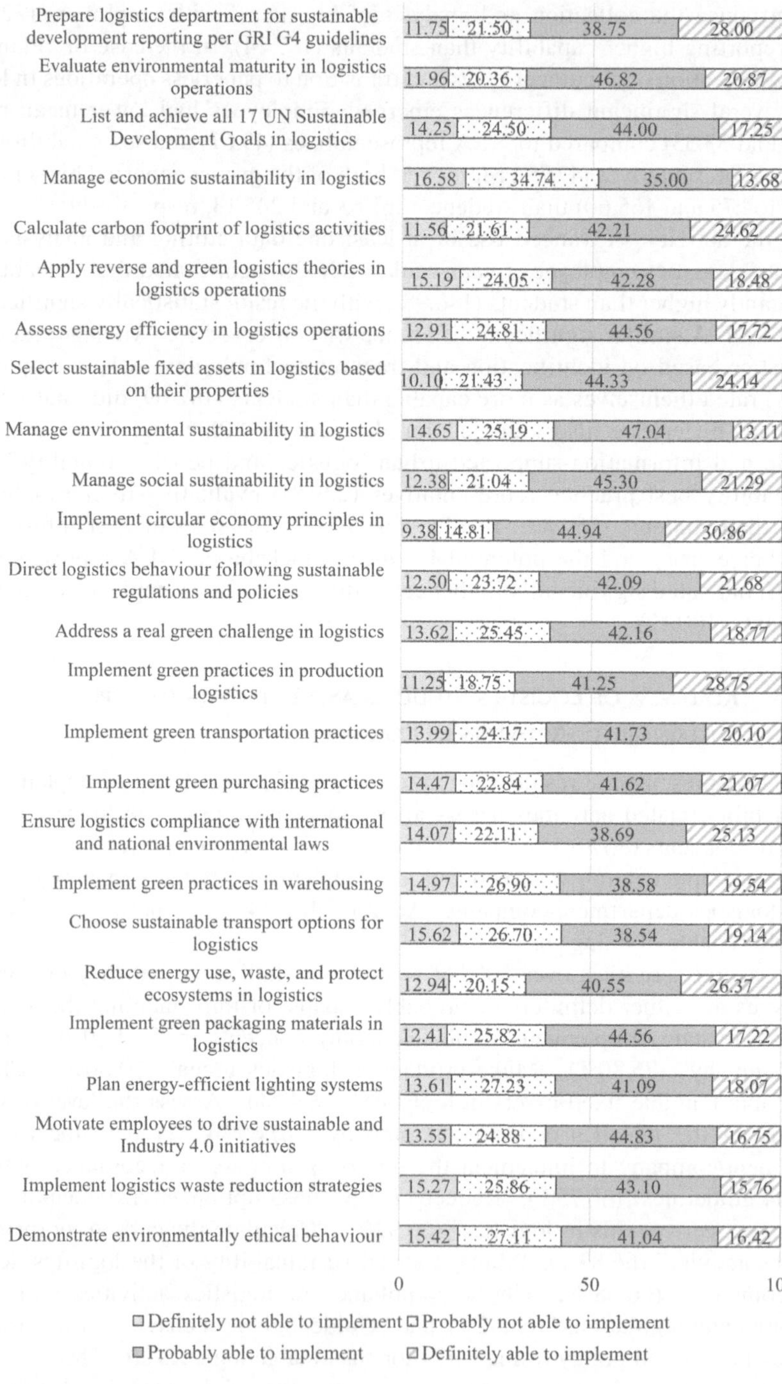

FIGURE 12.3 Readiness to implement activities related to sustainability in practice.

combined responses (definitely or probably not capable), this activity stands out, with 51.32% of respondents stating that they are not capable of implementing it. Other activities where more than 40% of respondents expressed a lack of capability include "Behave in accordance with environmental ethics" (42.54%), "Choose ecological solutions for internal and external transport" (42.32%), and "Make warehousing green" (41.88%).

Table 12.4 summarizes statistically significant differences in perceived capability to implement sustainability-related logistics activities, based on Mann–Whitney U tests.

TABLE 12.4
Perceived Capability Differences by Group (Mann–Whitney U Test)

Activity	Mean Ranks by Group	Mann–Whitney U Test
Manage economic sustainability of the logistics department/company	Employee (221.55) – Student (185.90)	7,857.00*
Manage environmental sustainability of the logistics department/company	NLA representative (254.75) – Student (163.33)	430.50*
	Sustainability best practice representative (230.06) – Student (164.41)	779.50*
Solve a real green challenge in the logistics sector	NLA representative (247.75) – Student (159.86)	430.50*
Introduce green packaging materials/packaging in logistics	NLA representative (260.13) – Student (162.81)	261.50*
Calculate the carbon footprint for logistics activity	Employee (216.54) – Student (182.89)	7,436.50*
Make warehousing green	Employee (214.98) – Student (183.57)	7,789.00*
Manage waste in the logistics department/company	Employee (220.37) – Student (183.68)	7,688.00*
Implement circular economy concepts in the logistics department/company	Employee (217.95) – Student (178.11)	7,142.00**
	Sustainability best practice representative (229.69) – Student (157.17)	670.50*
Motivate employees to introduce changes in order to achieve sustainable goals and I4.0 goals	Employee (226.43) – Student (185.73)	7,526.50**
Guide logistics behaviour in accordance with regulatory and policy frameworks for sustainable logistics	Employee (222.33) – Student (180.50)	7,086.00**
	Sustainability best practice representative (227.57) – Student (160.03)	640.00*
Asses the energetic efficiency of the logistics department/company	Employee (218.69) – Student (183.60)	7,529.00*

** The significance level is 0.01
* The significance level is 0.05

For the activity "Manage economic sustainability of the logistics department/company," employed respondents reported significantly higher capability (mean rank = 221.55) compared to students (185.90). Similarly, for "Manage environmental sustainability of the logistics department/company," NLA representatives (254.75) and those from companies with an outstanding sustainability track record (230.06) assessed logistics employees as significantly more capable than students assessed themselves (163.33 and 164.41, respectively).

Statistically significant differences were observed in activities "Solve a real green challenge" and "Introduce green packaging materials/packaging," where NLA representatives reported significantly higher mean ranks (247.75 and 260.13) compared to students (159.86 and 162.81, respectively). For more technical or process-oriented sustainability activities, such as "Calculate the carbon footprint for logistics activity," "Make warehousing green," and "Manage waste in the logistics department/company," employed respondents consistently perceived themselves as more capable than students, with statistically significant differences in mean ranks.

Significant differences also emerged for more complex activities. For example, in "Implement circular economy concepts" and "Motivate employees to introduce changes," employed respondents reported significantly higher perceived capabilities than students. A similar pattern was observed for the activity "Guide logistics behaviour in accordance with regulatory and policy frameworks for sustainable logistics," where both employed respondents (222.33) and representatives of an outstanding example of sustainability (227.57) reported higher perceived capability than students (180.50 and 160.03, respectively). Finally, for "Assess the energetic efficiency of the logistics department/company," employed respondents again showed a statistically significant advantage over students (218.69 vs. 183.60).

12.4 DISCUSSION

This study deliberately focuses on readiness to implement specific logistics activities in practice rather than on theoretical competencies. With the rapidly evolving nature of logistics activities, driven by Industry 4.0, digitalization, and sustainability, there is a pressing need to shift towards assessing practical skills that can be directly applied in the field. As Thai et al. (2024) suggest, competence is a multidimensional construct comprising knowledge, skills, and attitudes. However, in the context of the logistics industry's rapid transformation, it is the capability to effectively apply these skills in real-world scenarios that ultimately matters most.

Although some authors, such as Bandura (1997) and Pajares (1996), argue that individuals often overestimate their self-efficacy, the findings of this study suggest otherwise. Logistics employees rated their readiness to similarly to the evaluations provided by NLA and industry representatives. Contrary to expectations of potential overconfidence, the logistics employees in this study appear to have realistically assessed their capabilities.

The results directly address research questions stated in the methods section and are discussed in the following subsections.

12.4.1 Current Capabilities of Logistics Employees and Students in Industry 4.0, Digitalization, and Sustainability

Significant differences emerged in how respondents assessed capabilities related to specific Industry 4.0 activities. Out of the 16 listed activities, fewer than 50% of respondents indicated that they or logistics employees in general are probably or definitely capable of implementing four of them. The highest level of perceived capability was reported for "Understand the design and deployment of artificial intelligence based on knowledge of the theoretical background, real-life examples, and impacts of deployment," with 64.79% of respondents indicating capability. This was followed by "Implement optimization of logistics processes/routes/supply chain to achieve sustainability goals and use at least one optimization tool," at 63.97% of respondents assessing the capability to implement it. For the remaining activities, perceived capability ranged from 50.13% to 57.04% with an average of 54.30% of respondents indicating that they or logistics employees in general were definitely or probably capable of implementation.

Conversely, several activities were rated below 50% in terms of perceived capability, with respondents indicating that they or their employees were definitely or probably not capable of implementing them. The lowest-rated activity was "Introduce autonomous warehouse equipment," with only 46.08% indicating capability. Other activities with similarly low rating indicated:

- "Plan warehouse automation on a scale from partial automation to a dark warehouse based on economic feasibility,"
- "Decide on the possibility and appropriateness of using robots or collaborative robots to approach Industry 4.0/green/digitalization goals of the logistics department/company,"
- "Assess the level of Industry 4.0 maturity of the logistics department/company."

In the context of digitalization-related activities, all listed activities were rated by at least 50% as activities they or logistics employees in general are definitely or probably capable of implementing. On average, 62.70% of respondents rated themselves or logistics employees in general as definitely or probably able to implement. The highest-rated activity was "*Assess the level of digital maturity of the logistics department/company*," with 74.26% of respondents indicating capability.

Among the 25 sustainability-related activities, only one activity, "*Manage economic sustainability of the logistics department/company*," was rated as definitely or probably capable of implementation by fewer than 50% of respondents (48.68%). For all other sustainability-related activities, more than 50% of respondents indicated that they or logistics employees in general are capable of implementing them, with the average percentage being 63.38%. The activity "*Implement circular economy concepts in the logistics department/company*" received the highest rating, with 75.80% of respondents rating themselves as capable of implementation.

These findings provide valuable insights into the current readiness of students and employees in the logistics sector to implement key capabilities related to Industry 4.0, digitalization, and sustainability. While many digitalization and sustainability activities are already present to some extent, there is a clear need for further development, particularly in advanced Industry 4.0-related activities such as autonomous warehouse equipment, warehouse automation, and the use of robots. Moreover, although most respondents feel confident in managing digital maturity and sustainability concepts, targeted skill development will be necessary to fully prepare both students and employees for practical implementation within the logistics sector.

To answer the research question concerning which key capabilities related to digitalization, Industry 4.0, and sustainability are currently present among students and employees in the logistics sector, and which require further development to enhance readiness for real-world implementation, the findings suggest that while basic digital tools and certain sustainability activities are relatively well integrated, significant gaps persist in more advanced competencies. Given that a substantial proportion of respondents assessed themselves or logistics employees in general as not capable of implementing many of the listed activities, it is clear that targeted training programmes are needed to equip logistics professionals across all these critical areas.

Based on these findings, it is evident that logistics professionals must be better equipped with the skills and knowledge required to effectively manage Industry 4.0, digitalization, and sustainability activities in practice. This highlights the need for comprehensive training programmes that address all the identified gaps and prepare professionals to meet the evolving demands of the logistics industry.

12.4.2 Readiness Gaps: Comparing Self- and External Assessments of Logistics Capabilities

Significant differences were observed for 23 out of the 49 activities, including five related to Industry 4.0, eight to digitalization, and ten to sustainability. Comparisons were made between students and employees, as well as between students and the other respondent groups (employees, industry representatives, and NLA representatives). In addition, employees' responses were compared with those of the remaining three groups to identify further variations in perceived capabilities.

For activity "*Plan and lead digitalization of document flows and transition to paperless operations in logistics*," employees rated their capabilities lower than the representatives of NLA and sustainability best practices. For all other activities, no significant differences were observed between employees' self-assessments and external evaluations. This suggests that employees realistically assess their own capabilities.

For several activities, students rated their capabilities significantly lower than the representatives of NLA and leading companies, a difference observed in 18 comparisons across 15 listed activities. In addition, for 13 activities, students also rated their capabilities lower than employees did in their self-assessment. These findings suggest that students demonstrate a lower level of readiness to implement the identified activities in practice.

Several factors may explain the lower level of readiness observed among students. Firstly, unlike employees, students generally lack practical experience, as academic programmes tend to emphasize theoretical knowledge over hand-on application. In contrast, employees develop their capabilities through daily work and exposure to real-world challenges. This experience gap may lead students to underestimate their abilities, particularly those in relation to advanced technologies such as digitalization, Industry 4.0, and sustainability practices. Additionally, limited exposure to these technologies during their studies likely contributes to their lower self-assessed readiness.

In contrast, employees are generally more familiar with these technologies through their professional experience, which contributes to greater confidence in their capabilities. They also tend to have a more realistic understanding of own skills, grounded in the practical limitations and demands of their work. Students, lacking this experience, may overestimate or underestimate their abilities, often influenced by idealized academic scenarios. While students may focus primarily on theory, employees are guided by the pragmatic realities of day-to-day logistics operations. Psychological factors may also influence self-assessment: students, particularly those with limited hands-on exposure, may experience self-doubt, whereas employees often gain confidence after solving real-world challenges.

Finally, students and employees may perceive logistics tasks differently. Employees often have a broader understanding of the full scope of logistics activities, while students might focus on isolated tasks they have not yet performed in real-world setting. These findings highlight the need for increased practical exposure for students to be better prepare them for the demands of the logistics industry. To bridge the gap between theory and practice, educational programmes should incorporate more hands-on learning opportunities, such as simulations, case studies, and internships.

12.5 CONCLUSION

This study examined the readiness of logistics students and professionals to implement key activities essential for successful logistics transformation in the context of Industry 4.0, digitalization, and sustainability. While logistics professionals appear well prepared for basic operational tasks, gap remains in implementing advanced technologies and sustainable logistics practices. Interestingly, students rated their readiness lower than the current employees, further underscoring the need to strengthen education–industry links.

12.5.1 Theoretical Contributions

This study contributes to the existing literature by emphasizing the importance of the capability to implement advanced logistics practices rather than focusing only on knowledge acquisition. While previous research has highlighted the role of digitalization, sustainability, and Industry 4.0 in logistics, our findings suggest that implementing these practices remains a challenge.

12.5.2 Practical Implications

The results offer valuable insights for both industry and education, emphasizing the need for a greater focus on practical training and hands-on experience in logistics.

Companies are encouraged to invest in training programmes that go beyond theoretical knowledge, ensuring that employees gain the practical skills required to implement advanced logistics solutions. Furthermore, industry stakeholders should embrace lifelong learning strategies to keep pace with the rapidly changing logistics landscape. Regular training, certification programmes, and industry-led webinars can help employees continuously update their skills. Specialized training in areas such as sustainable logistics practices, artificial intelligence applications in supply chain management, or data-driven decision-making would support a more effective adoption of Industry 4.0 technologies.

Educational institutions and businesses should collaborate more closely to align curricula with evolving industry needs, ensuring that graduates are equipped with theoretical knowledge and the readiness to apply it in practice. This collaboration will help close the gap between education and the demands of modern logistics environments.

To effectively bridge the gap between classroom knowledge and real-world logistics challenges, institutions are recommended to implement practical, experience-based learning approaches. One example is the integration of simulation-based modules focusing on Industry 4.0 technologies – such as AI-driven warehouse management, digital twins, and Industrial Internet of Things (IIoT) systems – into logistics education. These simulations offer students the opportunity to explore complex systems and develop decision-making skills in a controlled, risk-free environment.

Additionally, establishing structured mentorship and internship programmes in collaboration with companies known for their excellence in sustainable logistics can provide students with valuable, first-hand exposure to current best practices. Such project-based learning experiences foster the development of both technical competencies and soft skills needed in a sustainability-driven logistics context.

By embedding these approaches into higher education programmes, institutions can better prepare graduates for the evolving demands of the logistics sector and ensure their readiness to contribute meaningfully from day one in the workplace.

12.5.3 Limitations and Future Research Directions

While this study significantly contributes to theory and practice, it has several limitations. Firstly, the study sample includes a high proportion of students compared to other respondent groups. This could lead to sample bias, as students' responses may reflect educational experiences rather than professional practice. Given that most students in the sample had limited professional experience, their perceptions might not fully align with the reality of logistics work. Consequently, the results may reflect more the state of logistics education rather than industry-wide trends.

Secondly, the study is geographically limited, including respondents from Austria, Italy, Poland, Portugal, Slovenia, and Turkey. This limited geographical scope may hinder the generalizability of the findings, particularly in regions with different educational systems, labour market conditions, or levels of technology adoption. Additionally, response bias should be considered, as respondents may have provided socially desirable responses rather than objective assessments of their capabilities or experiences.

The study's main limitation, however, lies in the reliance on self-reported data regarding experience and capability. Participants were asked to assess their capability to perform certain logistics activities, and representatives of outstanding industry and NLAs rated the capabilities of their employees. While this approach provides valuable insights, it carries the risk of overestimation or underestimation, as self-assessment may not always align with actual performance. For example, individuals may overestimate their capabilities due to unawareness of their limitations or underestimate them due to self-doubt.

Future studies could address these limitations by incorporating longitudinal studies and objective performance-based assessments. Longitudinal studies allow researchers to track skill development over time, providing a more dynamic understanding of how capabilities evolve with training and practical experience. Performance-based assessments, including practical simulations, observation methods, and third-party evaluations, would offer more accurate insights into actual competencies and help mitigate biases associated with self-assessment.

Expanding the scope of future studies to include professionals across various industries, experience levels, and global regions would further enrich the findings and offer a more robust understanding of how different contexts influence logistics capabilities in the era of digitalization and sustainability.

ACKNOWLEDGEMENTS

This research work is co-funded by the Erasmus+ project no: 2022-1-PL01-KA220-199 HED-000086366 (Sustainable Logistics 4.0: Digital and green skills for 200 boosting innovation and sustainability of the logistics sector).

REFERENCES

Bandura, A. (1997). *Self-Efficacy: The Exercise of Control* (Vol. 604). Freeman.

Cattaneo, L., Rossi, M., Negri, E., Powell, D., & Terzi, S. (2017). Lean thinking in the digital era. In J. Ríos, A. Bernard, A. Bouras, & S. Foufou (Eds.), *Product Lifecycle Management and the Industry of the Future. PLM 2017* (Vol. 517). Springer.

Cvahte Ojsteršek, T., Šinko, S., & Gajšek, B. (2023). Determining learning outcomes relevant for logistics higher education on sustainability and Industry 4.0. *Tehnički Glasnik, 17*(3), 447–454.

European Commission. (2007). Recommendation of the European Parliament and of the Council on key competences for lifelong learning. https://education.ec.europa.eu/focus-topics/improving-quality/key-competences

Gammelgaard, B. (2019). Editorial: Congratulations to IJLM on its first 30 years. *The International Journal of Logistics Management, 30*(1), 2–7. https://doi.org/10.1108/IJLM-02-2019-348

Gupta, A., Singh, R. K., & Gupta, S. (2022). Developing human resource for the digitization of logistics operations: Readiness index framework. *International Journal of Manpower, 43*(2), 355–379.

He, Y., Liu, W., Shi, X., & Gao, P. (2023). Teaching system transformation of logistics engineering major from the perspective of smart economy: An empirical study from China. *International Journal of Logistics Research and Applications, 26*(9), 1217–1240. https://doi.org/10.1080/13675567.2022.2038108

Islam, I. (2018). Automation and the future of employment: Implications for India. *South Asian Journal of Human Resources Management*, *5*(2), 234–243. https://doi.org/10.1177/2322093718802972

Khan, S. A. R., Jian, C., Zhang, Y., Golpîra, H., Kumar, A., & Sharif, A. (2019). Environmental, social and economic growth indicators spur logistics performance: From the perspective of South Asian Association for Regional Cooperation countries. *Journal of Cleaner Production*, *214*, 1011–1023. https://doi.org/10.1016/j.jclepro.2018.12.322

Kotzab, H., Teller, C., Bourlakis, M., & Wünsche, S. (2018). Key competences of logistics and SCM professionals – the lifelong learning perspective. *Supply Chain Management: An International Journal*, *23*(1), 50–64. https://doi.org/10.1108/SCM-02-2017-0079

Neumann, G., & Tomé, E. (2006). Knowledge management and logistics: Where we are and where we might go to. *LogForum*, *2*(3), 1–15.

Pacher, C., Woschank, M., Rauch, E., & Zunk, B. M. (2022). Systematic development of a competence profile for industrial logistics engineering education. *Procedia Computer Science*, *200*, 758–767. https://doi.org/10.1016/j.procs.2022.01.274

Pajares, F. (1996). Self-efficacy beliefs in academic settings. *Review of Educational Research*, *66*(4), 543–578.

Parham, S. (2018). The adaptation of the logistic industry to the Fourth Industrial Revolution: The role of human resource management. *Journal of Business Management*, *6*(2319).

Thai, V. V. (2012). Competency requirements for professionals in logistics and supply chain management. *International Journal of Logistics Research and Applications*, *15*(2), 109–126. https://doi.org/10.1080/13675567.2012.694859

Thai, V. V., Cahoon, S., & Tran, H. T. (2011). Skill requirements for logistics professionals: Findings and implications. *Asia Pacific Journal of Marketing and Logistics*, *23*(4), 553–574. https://doi.org/10.1108/13555851111165084

Thai, V. V., Tran, D. M., Nguyen, L. T. N., Nguyen, T. K., & Nguyen, T.-T. (2024). Revisiting the role of supply chain managerial competence in firm logistics performance: Do experience and education matter? *Sustainability*, *16*(23), 10459. https://doi.org/10.3390/su162310459

von Randow, M. (2008). Güterverkehr und Logistik als tragende Säule der Wirtschaft zukunftssicher gestalten. In *Das Beste der Logistik: Innovationen, Strategien, Umsetzungen* (pp. 47–53). Springer Berlin Heidelberg.

Winkelhaus, S., & Grosse, E. H. (2020). Logistics 4.0: A systematic review towards a new logistics system. *International Journal of Production Research*, *58*(1), 18–43. https://doi.org/10.1080/00207543.2019.1612964

Wong, W. P., & Tang, C. F. (2018). The major determinants of logistic performance in a global perspective: Evidence from panel data analysis. *International Journal of Logistics Research and Applications*, *21*(4), 431–443. https://doi.org/10.1080/13675567.2018.1438377

Zhang, Y., Montenegro-Marin, C. E., & Díaz, V. G. (2021). Holistic cognitive conflict chain management framework in supply chain management. *Environmental Impact Assessment Review*, *88*, 106564. https://doi.org/10.1016/j.eiar.2021.106564

13 Training Need Analysis of Digital and Green Skills for Boosting Innovation and Sustainability of the Logistics Sector

Beata Mrugalska

13.1 INTRODUCTION

Higher education institutions contribute to the United Nations Sustainable Development Goals (SDGs) due to their function and position within societies (Ketlhoilwe et al., 2019). They have a significant impact on the successful implementation of the three dimensions, including economic growth, social inclusion, and environmental protection. Referring to students' habits, they can support the goals oriented towards poverty, health issues, promoting well-being, gender equality, economic growth, responsible consumption and production, climate change, and peace and justice (Tilbury & Wortman, 2004). In order to achieve it, university curricula might include subjects such as systems thinking, foresight abilities, and comprehensive problem-solving. Interdisciplinary approaches should be central to both knowledge and action (Gual, 2019). By training, universities could help students understand how the SDGs can facilitate their future work. However, these goals should not only be included in international bachelor's and master's programmes. Universities could also be involved in new training programmes for non-university students. These students might participate in open online courses that contribute to lifelong learning programmes (i.e. Erasmus + programme) (Fumasoli & Rossi, 2021; Kasztelewicz et al., 2022). Furthermore, universities could adopt an ethical and transformative attitude and prepare all these individuals to address the pressing and significant challenges. It is essential for developed nations to ensure that education is accessible, affordable, and inclusive for everyone. Skilled students and professionals in developing countries should be encouraged to mobility (Gual, 2019).

Universities can also support the SDGs in their research, focusing on the expenses associated with their executing and the opportunity costs of inadequate investment. Action-oriented research, with the support of policymakers, the private sector, and civil society, will also be crucial (Alcántara-Rubio et al., 2022; Gual, 2019). As it is a universal framework, researchers can tackle the goals globally and

DOI: 10.1201/9781003611905-18

locally, emphasizing fundamental commonalities and differences between countries. However, in the literature we can mainly notice the development of specific research areas, such as environmental and/or circular sustainability (Gelmez et al., 2024; Umair et al., 2024b; Umair et al., 2025; Waqas et al., 2025), faculty training (Busquets et al., 2021; Kohl et al., 2022; Umair et al., 2024a), sustainable competences (Bianchi et al., 2022; Laasch et al., 2023; Wang et al., 2022), and the implementation of sustainability projects (Journeault et al., 2021; Willar et al., 2021).

According to Kestin et al. (2017), universities can incorporate the principles of the SDGs by their governance frameworks and operational policies, including those concerning employment, finance, campus services, support services, facilities, procurement, human resources, and student administration. It is suggested to enhance public involvement and participation in green initiatives. They should initiate and foster cross-sectoral dialogue, ensuring the representation of the higher education sector in national implementation efforts in activities focused on green policies and showing dedication to them. Universities can boost industry development and economic growth, and transfer knowledge, technology, and products (Pressman et al., 2019). While adopting experiential learning approaches, they can expand collaborations with industry, ensuring flexible access to green education (Akinsemolu & Onyeaka, 2025).

Business plays a vital role in the digital and green transformation. Companies implement standardized certifications and reporting frameworks, i.e. environmental, social, and governance (ESG) metrics, to communicate and monitor their sustainability initiatives efficiently (UNEP, 2024), which further highlights the necessity of digital skills in green employment (IDA Ireland, 2024). Thus, in order to implement it, modern environmental education should focus on innovative teaching methods and techniques that support society's needs, particularly those related to digital and environmental issues. Such notions should be embedded in the educational framework. According to Fomina (2020), implementing a research method of training helps students develop professional research thinking, providing them with a clear understanding of their core professional responsibilities and how to address them. It enhances their ability to implement modern technologies for gathering, processing, analysing, and organizing experimental data; introduces them to current research methodologies; and equips them to design research agendas. Additionally, it summarizes the importance of bibliographic efforts with the help of contemporary information technologies. The core of the research training method is to instruct students in scientific approaches for conducting laboratory or coursework, ultimately preparing them for their graduation qualifications.

This chapter discusses ongoing research being conducted to advance understanding of need of universities' engagement with the SDGs, especially focused on digital and green skills. The study investigates the alignment between university educational outcomes and industry requirements. In order to achieve it, surveys were conducted among the experts from universities and companies across logistics sectors. The obtained data allowed to propose the potential framework of a curriculum for higher education students. It included the topics and direct and indirect number of teaching hours.

13.2 METHODOLOGY

The presented study was implemented as a part of the Erasmus + project: Sustainable Logistics 4.0: Digital and green skills for boosting innovation and sustainability of the logistics sector. It included four universities: Politechnika Poznańska (Poland), Gaziantep Üniversitesi (Turkey), Universidade de Aveiro (Portugal), and University of Maribor (Slovenia), and three companies: Valuedo srl (Italy), ECQA (Austria), and Zerynth srl (Italy). However, for the purpose of this chapter, only data collected in Poland was presented.

The study consisted of two stages such as:

- determining key and essential topics for logistics higher education on sustainability and industry logistics 4.0,
- ranking of collected topics, and indirect and direct hours of teaching.

The first stage was the preliminary stage and provided data in the form of topics to further assessment using the Delphi method.

For the purpose of the study, the experts were selected on the basis of their experience and employment. They were representatives of two groups: academia and industry. The academics were affiliated in three universities in Poznań, Poland. The criterion of their choice was the fact that they had published at least two journal papers on logistics and/or sustainability in the last five years or taught a course from this field at a high school level in the last five years. In order to invite practitioners working in the areas of logistics and sustainability, personal contacts were used to access data. Following these criteria, the ranking data was obtained from seven and eight academics and four practitioners working in logistics, waste, and environment departments in two rounds of a Delphi method, respectively.

The questionnaires for the Delphi method were built on the basis of the preparation stage. It consisted of two aspects:

- importance of the topic/keywords,
- number of direct and indirect hours of teaching.

A 5-point Likert-type scale (not important at all – 1 score, essential – 5 score) was used to assess the participant's level of agreement. The first Google form was available for one month from 15 December 2022 till 14 January 2023. As soon as it was closed it, the topics were consolidated, and the experts to the second round were invited. A new Google form was available for eight days from 15 January 2023.

13.3 RESULTS AND DISCUSSION

The analysis of the first round of the Delphi method showed that most of the respondents paid attention to familiar subtopics. It resulted from the fact that the respondents had a specialization, i.e. transport, purchasing, supply logistics, warehousing, and stock. The final consolidated list of topics in the second round of national Delphi studies comprised 89 out of 147 topics (Table 13.1).

TABLE 13.1
Statistical Calculations for Topics

No.	Topic	M	SD
1.	Internet of Things	8	1.22
2.	Autonomous solutions in logistics (storage and transport). Autonomous vehicles and means of internal transport	9	1.48
3.	Digitalization of logistics processes	9	1.78
4.	Flexible manufacturing systems	8	1.86
5.	SAP system	9.5	1.08
6.	Database	9	1.56
7.	Computer simulations in logistics	8	1.70
8.	Virtual and augmented reality	8	1.24
9.	RFID	8	1.86
10.	Cobots in logistics	7	1.76
11.	Transport automation (automatic truck unloading systems)	7.5	2.41
12.	Big data in order to, for example, profile the image of the customer or predictive analyses to examine the supply chain in order to minimize threats	9	2.08
13.	Goods tracking systems	8.5	1.06
14.	Drones in logistics	6.5	1.93
15.	Information management. Construction of information models. Data analysis	9.5	1.56
16.	Mechatronics	7	1.57
17.	Communication. Digitalization of communication in forwarding processes	8	1.22
18.	Programming	7.5	2.56
19.	Automation and robotics	8	2.78
20.	Modelling of business processes	8	2.17
21.	Customer service	7.5	1.75
22.	Designing intelligent transport systems	8	1.44
23.	Regulations and standards for the storage of goods	7.5	1.62
24.	Statistical analysis	7.5	2.02
25.	Cybersecurity	8	2.27
26.	Calculation of the carbon footprint	5.5	2.94
27.	Environmental management and eco-innovations	7.5	1.83
28.	Sustainable production. Sustainable production resources	7	1.83
29.	Ecology	7	2.31
30.	Sustainable construction (e.g. warehouses)	6	2.18
31.	Green supply chains. Ecological transport (technical and organizational solutions) and storage (green warehouses)	6.5	2.19
32.	Materials science	6	1.71
33.	Commodity science	7	1.82
34.	Packaging	7.5	1.78
35.	Elements of chemistry	5	2.34
36.	Sustainable management of space and resources within intralogistics	7.5	1.19
37.	Integration of social, environmental, and economic issues as well as digitalization of processes and products	7	1.93
38.	SDGs in Poland and in the world	6	1.96

(Continued)

TABLE 13.1 *(Continued)*
Statistical Calculations for Topics

No.	Topic	M	SD
39.	Sharing economy (e.g. crowdsourcing)	5.5	1.92
40.	Climate changes	6	2.91
41.	Sustainable cities	5.5	2.42
42.	Sustainable development in business practice in terms of logistics (strategies, systems, and sustainable development indexes)	7	2.39
43.	Low-carbon economy and climate-neutral economy/logistics	6.5	1.51
44.	Preparing the organization to implement the reporting process in accordance with the GRI G4 guidelines	5.5	2.15
45.	Optimal use of raw materials	7	1.35
46.	Work-life balance	8	2.35
47.	Energetic efficiency	6	1.99
48.	Socially responsible investing	6.5	2.59
49.	Management based on asset protection. Reducing energy consumption, reducing environmental waste, and protecting ecosystems. Renewable energy sources in logistics	7.5	2.11
50.	Natural capital	5.5	2.32
51.	Creating and managing social partnerships	5.5	2.46
52.	Cultural, economic, and political forces influencing environmental attitudes and decision-making based on understanding science and technology	4.5	2.84
53.	Resource inventory	6	2.26
54.	Natural systems	5	2.15
55.	Recycling natural resources	6.5	1.62
56.	Electrical engineering, electronics, and electric drives	6	1.83
57.	Ecological solutions in internal and external transport	6	1.99
58.	Management of storage devices (equipment)	8	1.03
59.	Energy-saving lighting systems	6	3.03
60.	Directive 2008/98/EC, Standard BS 8001:2017, Standard TC 323, and Standard A/RES/70/1	6	2.41
61.	Logistic waste management system	6.5	1.97
62.	Numerical control of machines and devices	6.5	2.22
63.	Project management	7	2.02
64.	Digital economy concept	7	1.51
65.	Teamwork	7	1.75
66.	Circular economy. Return circuit parts and containers	7	2.47
67.	Triple bottom line	5	2.25
68.	Lean/agile organization management	6	2.19
69.	Evaluation and optimization of logistics processes	8	2.60
70.	User experience	7	2.43
71.	Multi-criteria support for decision-making processes in logistics	7	1.83
72.	Logistics planning	8	1.65
73.	Sustainable product	6	2.34
74.	Corporate social responsibility (ESG) risk analysis in logistics	6.5	2.05
75.	Use of electric vehicles	6	1.48

(Continued)

TABLE 13.1 ***(Continued)***
Statistical Calculations for Topics

No.	Topic	M	SD
76.	Corporate social responsibility	6.5	2.44
77.	Labour law and respect for employee rights	8	2.94
78.	Geopolitical knowledge	7	2.71
79.	Optimization of the use of means of transport to load weights in the logistics chain	7	1.98
80.	Management of production and warehouse facilities to extend the life cycle of machines	7.5	2.66
81.	New technologies in transport (renewable fuels and technical solutions)	7	2.09
82.	Intermodal transport	7	1.60
83.	Logistic centres	8	1.44
84.	Energy-sustainable warehouses and transshipment points	7	2.19
85.	Electronic documentation systems	8	2.27
86.	Quality control	9	1.56
87.	Production planning and optimization systems	8	2.61
88.	The needs of mobile application users, including the elderly and digitally excluded	6.5	2.99
89.	Application of spreadsheets in logistics	7.5	2.35

As it can be noticed the most often chosen topics were related to digitalization, Industry 4.0, and sustainability. In order to propose the curriculum in sustainable logistics 4.0, the final results are presented in Table 13.2, which includes the consolidated and merged topics into groups. The topics are categorized according to the study domains. Following the suggestions of experts, additionally, the area of "basic knowledge" was included to emphasize the need of extending general and basic knowledge in logistics.

On the basis of the study for each topic, we received information about direct and indirect teaching hours what is presented in Table 13.3.

TABLE 13.2
Consolidation of Topics

Topic Number	Proposed Title for a Group of Keywords/Topics
1, 4, 5, 6, 12, 15, 20, 25, 70, 85, 89	IT for logistics
7, 8	Simulations in logistics
2, 3, 9, 10, 11, 13, 14, 22, 64	Automation and digitalization of logistics processes
26, 27, 28, 29, 30, 31, 36, 37, 40, 41, 42, 43, 44, 45, 47, 49, 50, 54, 55, 57, 59, 60, 61, 66, 67, 73, 75, 81, 84	Sustainable logistics
68, 69, 72, 79, 80, 87	Optimization of logistics process
17, 38, 39, 46, 48, 51, 52, 65, 74, 76, 77, 78, 88	Social aspects in logistics
16, 18, 19, 24, 35, 62, 86	Basic prerequisite knowledge
21, 23, 58, 82, 83	Basic prerequisite knowledge in logistics

TABLE 13.3
List of Topics

	Keyword/Topic	Number of Direct Teaching Hours	Number of Indirect Teaching Hours
1.	Internet of Things	20.56	14.8
2.	Autonomous solutions in logistics (storage and transport). Autonomous vehicles and means of internal transport	25.40	19.20
3.	Digitalization of logistics processes	18.90	17.10
4.	Flexible manufacturing systems	16.22	13.78
5.	SAP system	25.80	21.90
6.	Database	16.33	19.89
7.	Computer simulations in logistics	15.33	23.89
8.	Virtual and augmented reality	11.63	11.67
9.	RFID	10.70	9.90
10.	Cobots in logistics	12.38	12.33
11.	Transport automation (automatic truck unloading systems)	13.67	14.44
12.	Big data in order to, for example, profile the image of the customer or predictive analyses to examine the supply chain in order to minimize threats	21.67	19.22
13.	Goods tracking systems	11.89	14.44
14.	Drones in logistics	9.22	7.44
15.	Information management. Construction of information models. Data analysis	17.33	21.11
16.	Mechatronics	21.22	16.11
17.	Communication. Digitalization of communication in forwarding processes	13.33	11.56
18.	Programming	29.44	34.44
19.	Automation and robotics	29.00	30.22
20.	Modelling of business processes	20.33	17.22
21.	Customer service	13.11	15.67
22.	Designing intelligent transport systems	19.22	17.67
23.	Regulations and standards for the storage of goods	12.89	11.33
24.	Statistical analysis	19.11	13.67
25.	Cybersecurity	20.33	17.89
26.	Calculation of the carbon footprint	8.00	5.00
27.	Environmental management and eco-innovations	10.25	8.75
28.	Sustainable production. Sustainable production resources	12.75	9.50
29.	Ecology	12.75	11.38
30.	Sustainable construction (e.g. warehouses)	13.13	9.75
31.	Green supply chains. Ecological transport (technical and organizational solutions) and storage (green warehouses)	11.38	11.00
32.	Materials science	15.13	12.63

(Continued)

TABLE 13.3 ***(Continued)***
List of Topics

	Keyword/Topic	Number of Direct Teaching Hours	Number of Indirect Teaching Hours
33.	Commodity science	19.00	13.63
34.	Packaging	15.38	9.75
35.	Elements of chemistry	21.63	14.25
36.	Sustainable management of space and resources within intralogistics	18.13	12.63
37.	Integration of social, environmental, and economic issues as well as digitalization of processes and products	15.25	10.63
38.	SDGs in Poland and in the world	13.17	11.67
39.	Sharing economy (e.g. crowdsourcing)	14.67	11.00
40.	Climate changes	8.67	5.00
41.	Sustainable cities	17.33	10.17
42.	Sustainable development in business practice in terms of logistics (strategies, systems, and sustainable development indexes)	14.50	12.50
43.	Low-carbon economy and climate-neutral economy/logistics	11.40	9.40
44.	Preparing the organization to implement the reporting process in accordance with the GRI G4 guidelines	10.20	11.40
45.	Optimal use of raw materials	15.50	10.83
46.	Work-life balance	16.29	13.57
47.	Energetic efficiency	26.50	16.67
48.	Socially responsible investing	14.20	11.00
49.	Management based on asset protection. Reducing energy consumption, reducing environmental waste, and protecting ecosystems. Renewable energy sources in logistics	15.00	13.60
50.	Natural capital	12.00	10.20
51.	Creating and managing social partnerships	9.50	7.00
52.	Cultural, economic, and political forces influencing environmental attitudes and decision-making based on understanding science and technology	25.60	10.00
53.	Resource inventory	12.00	8.83
54.	Natural systems	12.80	11.20
55.	Recycling natural resources	15.43	8.17
56.	Electrical engineering, electronics, and electric drives	21.00	14.00
57.	Ecological solutions in internal and external transport	14.67	9.60
58.	Management of storage devices (equipment)	15.57	15.83
59.	Energy-saving lighting systems	10.83	7.83
60.	Directive 2008/98/EC, Standard BS 8001:2017, Standard TC 323, and Standard A/RES/70/1	6.25	5.50
61.	Logistic waste management system	11.00	6.80

(Continued)

TABLE 13.3 *(Continued)*
List of Topics

	Keyword/Topic	Number of Direct Teaching Hours	Number of Indirect Teaching Hours
62.	Numerical control of machines and devices	15.17	9.40
63.	Project management	15.80	16.00
64.	Digital economy concept	13.40	8.20
65.	Teamwork	17.29	13.43
66.	Circular economy. Return circuit parts and containers	10.00	8.80
67.	Triple bottom line	9.00	9.00
68.	Lean/agile organization management	11.20	10.60
69.	Evaluation and optimization of logistics processes	15.20	10.20
70.	User experience	13.83	9.00
71.	Multi-criteria support for decision-making processes in logistics	12.20	10.00
72.	Logistics planning	26.50	15.00
73.	Sustainable product	12.17	8.17
74.	Corporate social responsibility (ESG) risk analysis in logistics	11.00	9.60
75.	Use of electric vehicles	10.80	6.00
76.	Corporate social responsibility	7.20	5.60
77.	Labour law and respect for employee rights	22.33	14.83
78.	Geopolitical knowledge	18.33	15.17
79.	Optimization of the use of means of transport to load weights in the logistics chain	10.20	10.00
80.	Management of production and warehouse facilities to extend the life cycle of machines	16.50	15.17
81.	New technologies in transport (renewable fuels and technical solutions)	19.17	18.50
82.	Intermodal transport	10.00	6.40
83.	Logistic centres	7.80	7.00
84.	Energy-sustainable warehouses and transshipment points	7.00	14.80
85.	Electronic documentation systems	9.00	9.40
86.	Quality control	13.20	7.6
87.	Production planning and optimization systems	15.80	18.60
88.	The needs of mobile application users, including the elderly and digitally excluded	10.80	7.00
89.	Application of spreadsheets in logistics	22.40	13.60

As it can be noticed, the perception of the need of teaching these topics was very diverse. Some of the subjects required 45 hours of teaching, whereas the others can be taught in 15 hours. These data allowed to elaborate the international level of the Delphi study and finally, it led us to defining curriculum (Cvahte Ojsteršek et al., 2023).

13.4 CONCLUSIONS

The European Union's strategic goals require business to integrate green practices and Industry 4.0 digital technologies to foster sustainability and competitiveness. Companies are encouraged to implement circular economy principles, leverage artificial intelligence (AI) and Internet of Things (IoT) for efficiency, and improve digital maturity to achieve these interconnected sustainability and digital ambitions. However, in order to achieve it, it is needed to engage the universities which could offer the curriculum for sustainable logistics 4.0 for higher education students. For this purpose, the research study was carried out where the main experts were academia and industry workers. They participated in the Delphi study which allowed to define topics for a potential curriculum of sustainable logistics 4.0. The respondents also suggested the number of hours which should be spend on studying them directly or indirectly. The proposed content does not only referred directly to sustainability and logistics, but it also indicated the need of teaching IT, automatics and robotics, management, and law. The respondents indicated many areas (i.e. material science and chemistry) not directly referring to logistics and sustainability which are necessary to be known. They may be included in the course or treated as a prerequisite requirement to participate in the course, for example, such as Slog4.0 course.

REFERENCES

Akinsemolu, A. A., & Onyeaka, H. (2025). The role of green education in achieving the sustainable development goals: A review. *Renewable and Sustainable Energy Reviews*, *210*, 115239.

Alcántara-Rubio, L., Valderrama-Hernández, R., Solís-Espallargas, C., & Ruiz-Morales, J. (2022). The implementation of the SDGs in universities: A systematic review. *Environmental Education Research*, *28*(11), 1585–1615.

Bianchi, G., Pisiotis, U., & Cabrera Giraldez, M. (2022). GreenComp – The European sustainability competence framework. In Y. Punie & M. Bacigalupo (Eds.), *EUR 30955 EN*. Publications Office of the European Union. https://doi.org/10.2760/13286

Busquets, P., Segalas, J., Gomera, A., Antúnez, M., Ruiz-Morales, J., Albareda-Tiana, S., & Miñano, R. (2021). Sustainability education in the Spanish higher education system: Faculty practice, concerns and needs. *Sustainability*, *13*(15), 8389.

Cvahte Ojsteršek, T., Šinko, S., & Gajšek, B. (2023). Determining learning outcomes relevant for logistics higher education on sustainability and Industry 4.0. *Tehnički Glasnik*, *17*(3), 447–454.

Fomina, N. V. (2020). Modern methodological approaches to environmental education at the university. *Journal of Physics: Conference Series*, *1691*(1), 012148.

Fumasoli, T., & Rossi, F. (2021). The role of higher education institutions in transnational networks for teaching and learning innovation: The case of the Erasmus + programme. *European Journal of Education*, *56*(2), 200–218.

Gelmez, E., Özceylan, E., & Mrugalska, B. (2024). The impact of green supply chain management on green innovation, environmental performance, and competitive advantage. *Sustainability*, *16*(22), 9757.

Gual, C. (2019). Are universities ready to have a real impact on achieving the sustainable development goals (SDGs). In J. Elías (Ed.), *Implementing the 2030 Agenda at Higher Education Institutions: Challenges and Responses* (pp. 41–43).

IDA (2004). IDA Labour Market Pulse Reveals Demand for Green Skills in Ireland Growing Twice the Rate of Global. https://www.idaireland.com/latest-news/press-release/ida-labour-market-pulse-reveals-demand-for-green-skills-in-ireland-growing-twice-the-rate-of-global Retrieved on 1st September 2024.

Journeault, M., Perron, A., & Vallières, L. (2021). The collaborative roles of stakeholders in supporting the adoption of sustainability in SMEs. *Journal of Environmental Management*, *287*, 112349.

Kasztelewicz, A., Tomaszewska, B., Rahner, S., Winter, I., & Voss, V. (2022). MOOC education for sustainable development. The Erasmus + project experience. *Polityka Energetyczna*, *25*.

Kestin, T. et al. (2017). Getting started with the SDGs in universities: A guide for universities, higher education institutions, and the academic sector. https://plattform-n.org/group/forum-plattform-n/file/university-sdg-guide_webpdf/download/University-SDG-Guide_web.pdf

Ketlhoilwe, M. J., Silo, N., & Velempini, K. (2019). Enhancing the roles and responsibilities of higher education institutions in implementing the sustainable development goals. In *Sustainable Development Goals and Institutions of Higher Education* (pp. 121–130). Springer.

Kohl, K., Hopkins, C., Barth, M., Michelsen, G., Dlouhá, J., Razak, D. A., Abidin Bin Sanusi, Z., & Toman, I. (2022). A whole-institution approach towards sustainability: A crucial aspect of higher education's individual and collective engagement with the SDGs and beyond. *International Journal of Sustainability in Higher Education*, *23*(2), 218–236.

Laasch, O., Moosmayer, D. C., & Antonacopoulou, E. P. (2023). The interdisciplinary responsible management competence framework: An integrative review of ethics, responsibility, and sustainability competences. *Journal of Business Ethics*, *187*(4), 733–757.

Pressman, L., Planting, M., Bond, J., Yuskavage, R., & Moylan, C. (2019). *The Economic Contribution of University/Nonprofit Inventions in the United States, 1996–2017.* Biotechnology Industry Organization & Association of University Technology Managers.

Tilbury, D., & Wortman, D. (2004). *Engaging People in Sustainability.* IUCN.

Umair, S., Waqas, U., & Mrugalska, B. (2024a). Cultivating sustainable environmental performance: The role of green talent management, transformational leadership, and employee engagement with green initiatives. *Work*, *78*(4), 1093–1105.

Umair, S., Waqas, U., Mrugalska, B., & Bystrov, I. (2024b). Linking transformational leadership to green human resource management (GHRM) and employee engagement in green initiatives. *International Journal of Sustainable Engineering*, *17*(1), 1083–1095.

Umair, S., Waqas, U., Mrugalska, B., & Durrah, O. (2025). Green human resource management, green environmental passion, pro-environmental behaviour, and sustainable environmental performance. A sequential mediation analysis. *Economics and Environment*, 95(4), 927–927. https://doi.org/10.34659/eis.2025.95.4.927

UNEP (2024). We are all in this together - Annual Report 2024. https://www.unep.org/resources/annual-report-2024 (Retrieved on 1st August, 2025).

Wang, Y., Sommier, M., & Vasques, A. (2022). Sustainability education at higher education institutions: Pedagogies and students' competences. *International Journal of Sustainability in Higher Education*, *23*(8), 174–193.

Waqas, U., Umair, S., Mrugalska, B., & Ali Sulaiman, M. A. B. (2025). Twin transformation: The role of digitalization with artificial intelligence, big data and circular supply chain in circular economy performance for sustainability. *Equilibrium. Quarterly Journal of Economics and Economic Policy*, *20*(2), 613–645. https://doi.org/10.24136/eq.3350.

Willar, D., Waney, E. V. Y., Pangemanan, D. D. G., & Mait, R. E. G. (2021). Sustainable construction practices in the execution of infrastructure projects: The extent of implementation. *Smart and Sustainable Built Environment*, *10*(1), 106–124.

Part VI

Education Institutions and Sustainability

14 Digitalization as a Catalyst for Sustainability in HEIs

Muhammad Zulkifl Hasan, Muhammad Zunnurain Hussain, Sonia Umair, and Umair Waqas

14.1 INTRODUCTION

14.1.1 Background and Context

The technology and sustainable needs, which connect at an unprecedented level, have major transformation on the educational system at present. As part of this transformation, digital technology advanced all stages of knowledge development as well as the distribution procedures along with university social duties (Akour & Alenezi, 2022) As a result, the physics now goes beyond the limits of the classroom as the higher education institutions (HEIs) function as an interconnected dynamic ecosystem within the virtual and physical domains. At the same time it is a solution to broader social movements, it fuels further change on the information economies and labor markets as well as the shifting of global economics (Gorina et al., 2023).

When digital transformation became a priority, the institutions of the 21st century required to become sustainable. Today, HEIs are acting as major forces in sustainable development due to environmental deterioration and climate changes and growing social and economic instability (Umair et al., 2024; Waqas et al., 2025) and also play an important role in knowledge distribution. To demonstrate sustainability, contemporary universities' operations and the instruction of sustainable grads should seek to investigate international environment and social calamities.

Digitalization opens new ways and means of sustainable values in transformation. Beyond its offering of technological changes, the digital transformation is more an operational transformation that guarantees support to the sustainable approach at all institutional levels. There is a way to save energy, money, and reduce operational costs by integrating cloud computing with big data analytics and artificial intelligence (AI) and virtual learning capabilities with HEIs. In line with the use of the digital tools, sustainability elements follow sustainable digital systems and substance for equal access to technology and an ecofriendly manner to IT operation and medical manner of the technological being.

As such, the following obstacles stand in the way of the adoption of these types of platforms that have sustainability targets associated with their installation. Through interaction with cultural elements the implementation course is set as policy regulations emerge from the interaction with these capacities in conjunction with

DOI: 10.1201/9781003611905-20

availability of resource. Sustainable digital transformation is a process of the daily solution of different problems and constant debates on nature and development itself. For the very same reason, HEIs need to understand profound digitalization knowledge at depth to generate a societal development toward sustainability through the reshaping of its educational system.

14.1.2 Purpose and Scope of the Chapter

This section reviews multiple ways digital technology improves sustainability results at colleges and universities. The research studies how to properly use digital technology to strengthen both environmental protection and social justice while handling wealth successfully. Our method shows how digitalization serves both its technical base and its institutional purpose while creating social value. Our main goal is not to simply promote digital tools but to disclose their benefits for sustainability and show how best to use them for universal success.

Our study uses strategic principles to build its theories and demonstrates clear application scenarios. Our review starts with explaining the base principles that combine digital technology and sustainability in educational institutions. The chapter provides real examples from global institutions that show these strategies operating together in actual workplaces. The text explains why digital transformation challenges continue with examples that combine limited internet access and expensive prices with school opposition and planet damage concerns. The author supplies specific advice and rules for university executives and officials to manage emerging technology applications. This chapter develops a complete solution to upgrade higher education discussions on digital learning and sustainable practices. HEIs take on two tasks by serving both as technology implementers and as leaders to shape international social development.

14.1.3 Structure of the Chapter

This topic receives organized analysis through eight connected sections in the chapter. This section examines digitalization since applications of modern technological elements in higher education. HEIs achieve sustainable goals through implementing objectives from the triple bottom line (TBL) and SDG frameworks according to Section 14.3. Section 14.4 of our investigation merges previously established theories to establish how digital instruments support sustainable functions combined with institutional coordination of both agendas. Section 14.5 demonstrates quantitative research validating that digitalization enhances both environmental safeguarding efforts and financial health improvements and boosted social programs within the higher education sector. Section 14.6 demonstrates successful practices resulting from all nations with actionable success cases coming from developed and developing countries. The concluding section explores upcoming scenarios. The last section analyses future technology trends while recommending essential policy requirements together with research needs. Section 14.8 concludes the chapter by identifying essential research outcomes and postulates about upcoming transformations in higher education due to digital-sustainability interdependencies.

14.2 UNDERSTANDING DIGITALIZATION IN HIGHER EDUCATION

14.2.1 Definition and Evolution of Digitalization

An entire university system must undergo full technological modernization to affect its entire academic operations from educational teaching to scholarly development to research support services. The definition of information and communication technologies (ICTs) in higher education sector cannot rely solely on technological deployment or user convenience systems. Higher education digitalization works as a vital institutional transformation that modifies knowledge development frameworks and learning approaches while changing how universities handle student assets. During the final two decades of the 20th century modern educational methods expired because computing technology ascended to its peak leading to mobile devices and internet-based technology creating a modern revolution. Digitalization established creating digital duplicates as its essential target from physical materials. The initial digital period featured both activities of material digitization and systematic record system construction with email platform implementation. Educational software along with operational systems relied on first digital engagement model technology to create their administrative robotization tools.

The rapid development of digital transformation remains quick due to persistent growth of AI and machine learning and blockchain and Internet of Things (IoT) technologies worldwide. University education allows students to create customized educational pathways through the combination of operational efficiency advancements and digital facility monitoring information. Educational organizations must merge traditional classrooms with digital education because quick digital progress requires active student-led learning instead of passive teaching methods. The school transformation process requires multiple organizational elements to change as performance outcomes across the institution remain unstable (Perrin & Wang, 2021) An institution's readiness emerges because it possesses different digitalization capabilities which achieve preparedness through effective infrastructure and financial backing together with staff readiness to employ the system. Digitalization merges technological developments and socio-institutional procedures which apply vision to development initiatives based on context.

14.2.2 Key Digital Technologies in HEIs

HEIs depend on digital technology systems as fundamental transformation elements which help improve academic excellence and operational success. Distance learning education functions through Learning Management Systems that include Moodle and Blackboard and Canvas for delivering both instant and delayed teaching using digital feedback systems. The extensive data operations of educational organizations become protected when they use cloud computing services while their expanded service offerings need minimal physical building growth.

Large data analysis helps educational institutions develop superior management solutions that gain research-based support and student assistance through data interpretations. Modern controllers within the field of modern technology develop virtual reality and augmented reality programs for new learning experiences but these programs presently focus primarily on medical education and engineering practice and artistic domains. The blockchain technology helps establish damage-proof credentials to operate effectively on clear documentation platforms. Computing operation of smart facilities requires IoT technologies to connect infrastructure management systems with digital networks for control of security components as well as temperature control and illumination systems and resident monitoring platforms. The individual operation of each technological system occurs without dependency on alternative systems. The strategic alliance of current technologies produces advanced operational networks that deliver superior value through excellent learning results together with organizational success alongside sustainable environmental outcomes. The path to organizational success becomes clearer when leaders show active backing of technological change combined with employee training about new platforms across maintained evaluation systems.

14.2.3 Digital Transformation and Digital Adoption

People are confused about educational advancement because of the combination of digital system implementation and the digital transformation processes. Using digital tools in relation to one's workplace operations enables an adoption of digital practices. The random clustering approaches make it possible for the institutions to perform digital transformations, however, they require some extra capabilities to design the rest of the operational aspects. They provide instant benefits when building new processes, and there is no need to implement change in the operational model or in the structure of the organization.

The digital transformation turns into a powerful process of organizational transformation that results in the complete reorganization of the learning function to bring together strategic objectives and to incorporate digital innovations. New tools are introduced to organizations that alter employee values and organizational relationships as well as implementation procedures. In order to achieve the successful transformation outcomes, continuous innovation work on HEIs should establish continuous innovation work on flexible operational systems and education sustainability learning. Because both university governance models and the processes of educational restructuring have constituted changes, students have to be involved in the development of educational learning experiences (Ben Kasmia & M'hamed, 2023)

Inclusive participatory transformation can be successfully endorsed with leaders who clearly understand and implement inclusive participatory methods. Transforming is realized only though the linkage of existing principles to technology systems, so that this requires vision at the highest level of implementation of technology in educational institutions. Each digital transformation is given the ability in ethical leadership to provide new solutions that enable privacy and promote equity and environmentally friendly accessibility. Therefore, while the HEIs have necessitated the implementation of fundamental changes rather than the complete structurization of their system, their main operational objective in this period is to identify important changes. There is both downloaded digital technology in all universities, as well as in full digital transformation, but digital transformation can be said to have given universities the ability to control complexity while simultaneously allowing for success in the practices of sustainability.

14.3 CONCEPTUALIZING SUSTAINABILITY IN HIGHER EDUCATION INSTITUTIONS

14.3.1 The Triple Bottom Line: Environmental, Social, and Economic Sustainability

The environmental foundation of sustainability now extends its protection to all economic and social development areas. Sustainability as a principle gives educational activities at colleges and universities above and beyond its initial requirements for added responsibility and governance. The idea advanced into an essential foundation that demonstrates institutional recognition while establishing direction and establishing accountability. Educational institutions at higher levels must be leaders in sustainable practices because they must show strong expertise and leadership in sustainability through their intellectual and social duties (Umair et al., 2024). Several HEIs utilize the TBL as their fundamental model approach to implementing sustainable practices. John Elkington developed the TBL which serves as a tool to evaluate institutional performance by combining financial data with environmental as well as social accountability responsibilities. HEIs can evaluate their entire operations and funding choices through the conceptual format which includes analysis of all institutional-level effects.

Higher education institutions work toward environmental sustainability through sustainable power transition and pollution reduction efforts and natural resource management and space development for environmental friendliness. The attainment of sustainability targets by HEIs involves the creation of facilities with energy-saving capabilities as well as the provision of green transit along with the digitalization of printed operations. The institution succeeds in reaching ecological targets because its commitments serve both environmental protection and sustainability education on campus grounds. Through social sustainability efforts universities seek to establish open communities that ensure everyone receives equal opportunities to maintain their well-being. HEIs undertake teaching responsibilities as they maintain independent status as a social community. The realization of social sustainability within universities requires providing opportunities for education fairness to students who represent various gender groups from distinct economic regions and geographic locations alongside provisions for mental health care services which foster citizenship development alongside cultural diversity acceptance. Official documents should display institutional values precisely as stated but these principles should exist fully in instructional activities together with staff hiring processes and security operations and student support initiatives.

HEIs require sustainable financial operations since they receive funds from inconsistent grants and increasing student debts and escalating operational expenses. All institutions providing higher education must create financial stability as they maintain exceptional academic standards. Several factors contribute to higher education sustainability including budgetary experience backed by resource management competence and additional approaches for revenue generation combined with affordable service delivery. HEIs reach economic wellness by sustaining economic sustainability which lets them maintain student-focused financial stability and value preservation. Institutions which use TBL achieve better sustainability integration through daily development of strategic alliances between the three pillars.

14.3.2 Sustainability Frameworks in Higher Education

During the past two decades educational institutions which supported sustainability-issued international declarations to create institutional frameworks. The established protocols enable universities to track their performance through public goal setting without exceptions. The United Nations proclaimed Sustainable Development Programs as the most extensive framework after they passed their adoption in 2015. The 17 SDGs operate as connected frameworks to tackle global issues which include poverty and educational inconsistencies and environmental destruction and climate change problems. HEIs undertake worldwide orders from the SDGs to develop sustainable contributions by focusing on policy research development while executing operations and creating curriculum and programs. University institutions focus heavily on educational targets under Goal 4 because they pursue simultaneous attainment of targets from Goals 5, 7, 11, and 13 and their respective linked objectives (Leal Filho et al., 2024).

Several associated frameworks give support to the SDGs by providing specialized educational resources to address particular needs. Higher education leaders established the Tallies Declaration as their inaugural environmental sustainability promise that remains a key historical declaration for sustainability efforts. An operational framework comprises three fundamental responsibilities which institutions must implement for environmental education as well as sustainable operations and community involvement success. HEIs can assess their performance across essential domains through Sustainability Tracking, Assessment & Rating System (STARS) which the Association for the Advancement of Sustainability in Higher Education maintains.

14.3.3 Global Goals and Institutional Commitments

HEIs need to prove their sustainability goals through concrete action plans because societal concerns about sustainability and environmental standards increase. The organizations which obtain financial support from funding agencies and governmental institutions and student/alumni donors need to demonstrate their outcomes in a transparent manner. Universities worldwide develop sustainable systems to fulfill stakeholder demands through declaration coordination with planning infrastructure and dedicated units that handle reporting. Universities establish sustainability offices as exhibitions of educational commitment by establishing organizational units that formulate sustainable campus strategies for resource management. The offices consolidate departmental resources to establish harmonious joint operations between supply chain principles and waste management and sustainable content production together with community engagement programs.

Chief Sustainability Officers attain their advanced sustainability roles by serving as leaders at university governance through executive positions.

Educational institutions have integrated sustainability principles into their educational transformations so that all subjects inside their curriculum are influenced. The integration of environmental studies and sustainability science with green technology and social innovation operates in thousands of academic institutions throughout their educational departments. Teaching personnel at the organization must incorporate sustainability content throughout core subjects as well as specific courses while preparing students for worldwide problem-solving. Laboratories and research centers serve as cluster programs for university environmental development to focus on sustainability-related problems and energy-efficient climate adaptation while advancing city circular economy practices. Public educational institutions participate with both government entities and civil society organizations and private entities in order to establish civic institutions and maintain their role in community development efforts. University institutions maintain power through their dedication to both the Global Alliance for Sustainability and Sustainable Development Solutions Network and regional sustainability consortia. Through their networks education institutions can both exchange knowledge and team up for sustainability projects and execute advocacy initiatives regarding policy reforms and teaching practices. Sustainability initiatives gain support from digitalization because it functions both as an effective tool while also minimizing time consumption. Digital platforms at universities enhance sustainability programs by delivering educational resources with tracking functions and virtual networks which provide premium educational opportunities to any student. The section displays how digitalization enables crucial sustainable developments which universities need to execute in their academic practice (Ramaswamy et al., 2021).

14.4 INTERSECTION OF DIGITALIZATION AND SUSTAINABILITY

14.4.1 Digitalization as an Enabler of Sustainable Practices

With the availability of new digitalization methods and instruments in the coordinated transformation of university systems, HEIs undergo a meaningful radical change of the operations structure. Organization's use of core digitalization plan leads to institutional sustainability by advancing on environmentally friendly and socially sound and economically sound operations. The objective of the tool is to speed up the implementation of sustainability within principal frameworks of organizational structures of HEIs (Robertsone & Lapiņa, 2023) Under such parameters, lower environmental resource requirements are also protracted in digital technologies which further allow institutions to have better resource control. By using cloud-based systems, universities will lower the storage hardware costs in the storage, and thus lower the energy consumption of their facility servers. Educational institutions, IoT smart projects enable the institution to monitor the utility in real-time and detect resource inefficiencies as a result of avoiding cost. This method ensures that the implementation of electronic document platform and online communication system enables the educational institutions to adopt low carbon operation model.

Part of social sustainability programs, students from the disadvantaged backgrounds get equal access to education through digital revolution. The system wipes out three essential barriers, which inherently exist to keep universal opportunities for higher-level studies out of reach due to financial conditions or geographic position, through its educational tools. AI assistance to the assistive technologies facilitates development of personalized educational scenario for promoting results of the learners suffering from learning disabilities. Universities create new digital communication networks on their platform to engage students socially and to share knowledge to strengthen the bonds with its entire student group (Mothafar et al., 2024).

Digital procedures make the operation of educational organizations more efficient in reducing the costs of expenses of the operation and this frees the saved funds for sustainable end use. Digital operations replace manual labor processes reducing labor, which universities are able to implement systems of automation that leads to development of flexibility. Universities use the financial savings generated by education operations to set up sustainable infrastructure, take part in environmental studies, establish programs to support student well-being, and so on. Digitalization is necessary for the acceleration of sustainability because the automated institutions are in a competitive position to perform better due to increased operational visibility and rapidity of responses.

14.4.2 Strategic Integration: Policies and Institutional Digital Strategies

HEIs need to transform their random technology use into perfectly aligned digital planning strategies that serve sustainability targets for sustainable digital growth (Anh et al., 2024). Every section of organizational departments that includes leadership and education and infrastructure needs policies to unite digital innovation goals with sustainability targets. Major research institutions have incorporated parallel integration points between sustainability targets and digital structures into their digital management systems and sustainability operations platforms. Economic social capability assessment and environmental sustainability review are applied to college campuses during the assessment process of complete digital initiatives. There are several institutions in which the CIO and CSO are enabled to work together in the development of frameworks which integrate technology governance practices with sustainability requirements.

Different establishments have to go ahead and devise certain benchmarks to gauge the sustainability outcomes from their digital projects. Modern institutions now have real-time data to track its energy conservation achievements as well as emission reduction as well as to access equity performance capabilities through digital dashboards to make real-time data-based choices at immediate. With a new digital transformation, these responsible standards of IT infrastructure acquisitions are an important part of the plan, since they allow for responsible data management and digital equality reduction. To develop institutional digital strategy, it is needed to integrate fundamental digital principles and sustainability issues as well as concerning moral views to cover digitalization. Organizational priorities include making accordant safeguards for data privacy and algorithm bias; the organization actively rejects seeing technology as a solution for long-lasting sustainability problems. Such higher education digital strategies are therefore crucial for providing sustainable development with a bridge between lively progressiveness and relevant ethical requirements and principles of societal equity.

14.4.3 Challenges and Barriers to Sustainable Digital Transformation

However, there are different barriers organizations face to achieve sustainable digital transformation benefits that digitalization provides to sustainability actions. If, on the other hand, care is not taken to remove these organizational and technological obstructions, and financial and cultural elements are paid attention to, advancement is unlikely to occur. The digital gap is still crucial as it impacts institutions in their various segments and groups of institutions. Modern digital systems and a skilled workforce are provided to educational institutions in resource rich areas of developed regions from the technological base, but with such an advantage institution in low- and the middle-income areas with weak digital infrastructure and weak setup in information technology do not enjoy. Educational institutions have differential distribution of digital tools thus extending student socioeconomic and geographic disadvantages. The major stumbling block of financial barriers are the first that the students and the organizations face. Digital technology acquisition requirements have initial high costs and future financial productivity growth. During the entire hardware acquisition process and during software development, as well as training and operational support of the instructor to institutions, budget cuts in public universities actually worsen the financial challenges. Strategic foresight during digital investments leads to misalignment of sustainability goals resulting diversity and duplicity of broken technological systems that fail to realize the intended transformative goals.

Besides existing work procedures coupled with slow bureaucratic processes and employees' reluctance to change, digital transformation is inherently hostile toward institutions as it represents a change to how things traditionally have been done. Educators and workers in the workforce that lack digital skills in either educational activities and operation settings, as well as critical teaching methodologies hinder them from using novel educative strategies in those processes. Integration, however, does not work in a situation where change initiatives are not effective in developing the staff hence superficial implementation as opposed to comprehensive integration. The environmental consequences of the digitalization of the world experienced by researchers are too important to be left out in detail. Three key operational challenges are, however, raised by the use of digital tools in their production of e-waste, their electrical power requirements and their reliance on extensive supply chains with rare earth materials, and the emissions caused by them. Haphazard digital infrastructure management during its entire lifespan and culminating with waste disposal operations will have a negative effect on several sustainability objectives.

The integrity challenge of digital sustainability is that organizations have to deal with ethical problems of data privacy violation and biased algorithms. To address the problems posed by digital resources, institutions must embark upon open management and democratic oversight as well as complete accountability. According to the various challenges, it is possible that full equal digital transformation depends on support in money. This method is directly applied and thrusts in major transformational results that can be used over and over again in many educational institutions.

14.5 DIGITALIZATION'S IMPACT ON SUSTAINABLE PERFORMANCE

14.5.1 Environmental Impacts

Institutions that employ digital methods produce substantial environmental impacts which affect both their operational facilities and their energy usage and environmental impact. The current digital solutions act as fundamental operational tools since educational institutions need solutions to meet internationally set climate change targets. Because universities adopted digital learning platforms with administrative paperless systems, they succeeded in decreasing their material usage on a global scale. The integration of electronic assignments and cloud-based online testing alongside book storage systems allowed universities to reduce their paper consumption which led to decreased forest depletion, as shown in Figure 14.1.

Digital technologies have twofold capabilities to save energy expenses while simultaneously making energy usage more efficient. Universities achieve detailed electricity detection through their implementation of IoT sensors and automated climate controllers operated with real-time software. The integration of automation systems powered by technology regulates heating and cooling together with lighting to minimize total power usage while decreasing environmental impact. The energy efficiency of smart buildings depends on their ability to track occupancy frequencies to

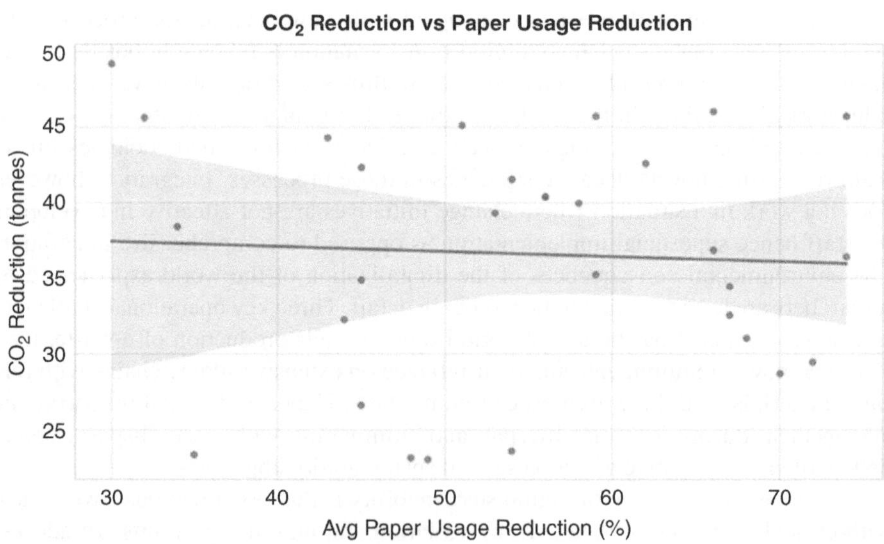

FIGURE 14.1 CO_2 reduction.

Source: Author's own analysis.

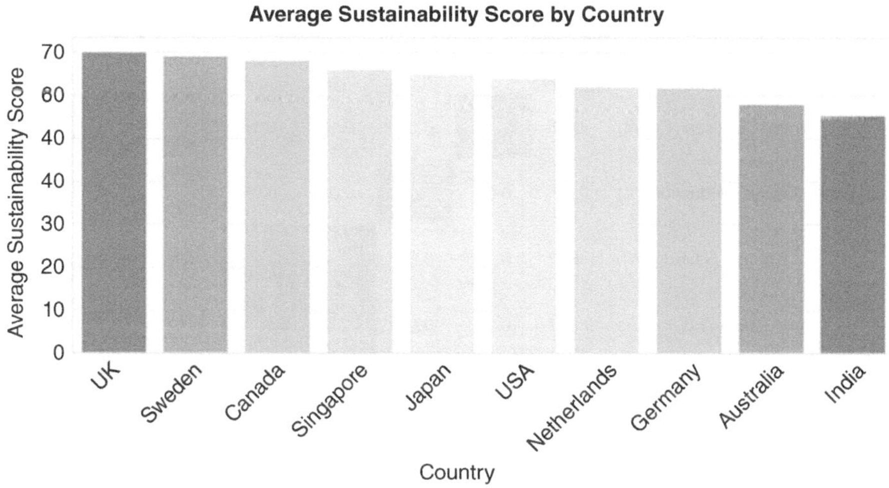

FIGURE 14.2 Average sustainability score.

Source: Author's own analysis.

cut down electricity usage and air conditioning needs proportional to actual physical building occupancy. The control method enables institutions to achieve dual advantages by reducing both power expenses and environmental effects at the same time.

In order to cut down the resource cost to perform academic research on future educational facilities through environmental data analysis with a building design testing, computer-based simulation models are needed. The environmental science degrees using Geographic Information Systems and spatial technology provide better learning experiences to the students but need to replace the conventional field study and laboratory activities. There is no environmental neutrality in the basic state of digitalization. Electronic garbage is required digital hardware production that both produces further waste and requires maintenance support and utilization of such tremendous energy of non-renewable energy. In other words, that sustainability must be practiced from up to down; that is, HEIs must invest in green technology while developing environmentally sustainable data centers and waste management system. Because HEIs manage their decisions on technology for institutional alter or convenience features, rather than merely for cost savings, they desire to talk about the environmental outcomes of the digital projects as can be seen in Figure 14.2.

14.5.2 Economic Efficiency and Cost Reduction

Most universities, as its performance outcomes are reached, rely heavily on the use of current economic sustainability to do so as it operates within a high level of financial instability and fluctuating enrolment numbers, and financial limits. The operational process of digital transformation by the universities is a basic process aimed toward enhancing the process efficiency by multiple institutions through different operational activities which in turns lessen operational costs to economize saved

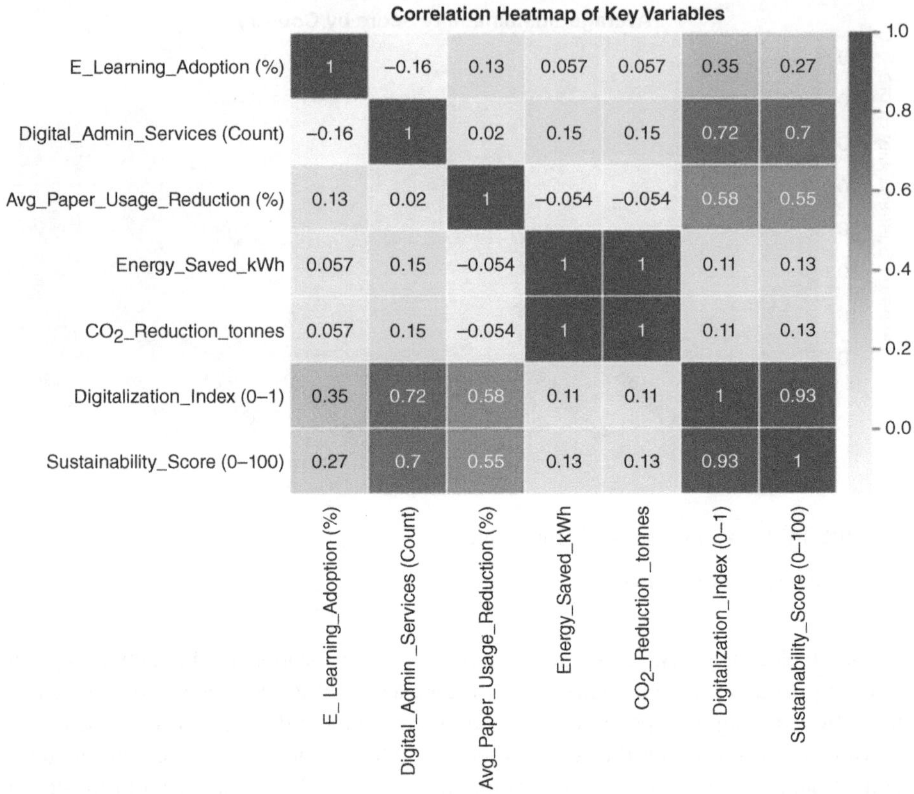

FIGURE 14.3 Correlation heatmap.

Source: Author's own analysis.

funds toward the strategic focus areas as illustrated in Figure 14.3. Through such a transformative process students as well as public get improved financial returns to what is simple offering in monetary saving possibilities. Via digitalization of administrative work, HEIs are found to increase the accuracy rate and shorten the time invested by students and employees in processing in order to reduce the workload of the working staff. They require the operational savings of digital admission systems that make students accomplish online admission and payment of tuition and automate grading processes. In this regard, implementation of cloud computing provides the educational institutions with some additional IT resources, which do not need enough new server and storage facilities. This method reduces academic institution's capital requirements as well as minimizes its operating expenses while yielding also flexible service expansion capabilities as shown in Figure 14.4. Educational services also show a great quantity of financial expansion in the delivery practices. To expand into a bigger student population beyond previous enrolment boundaries without the need of high facility needs, institutions employ hybrid and online learning as illustrated in Figure 14.5.

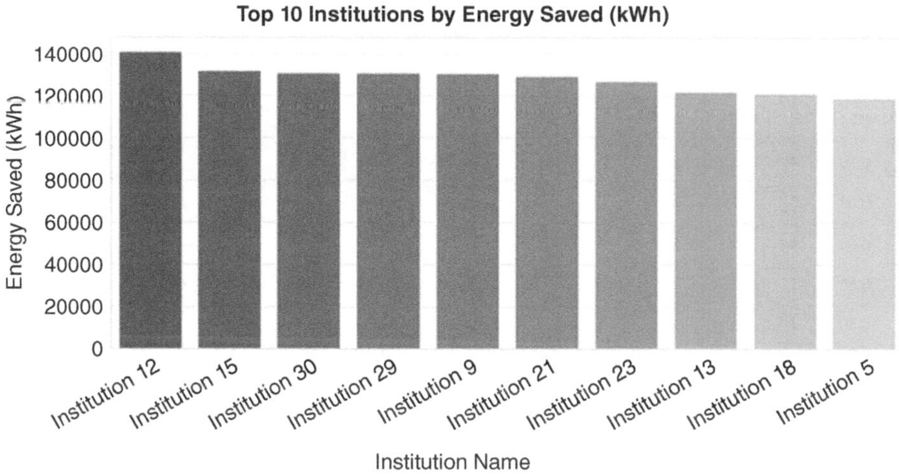

FIGURE 14.4 Top 10 distribution.

Source: Author's own analysis.

Virtual classroom technology provides universities with an opportunity to develop reduced cost remote programs with supporting admission growth of international students. Learning platforms allow students to use open educational resources (OERs), which saves money on textbooks and gives teachers new technologies to present the curriculum content.

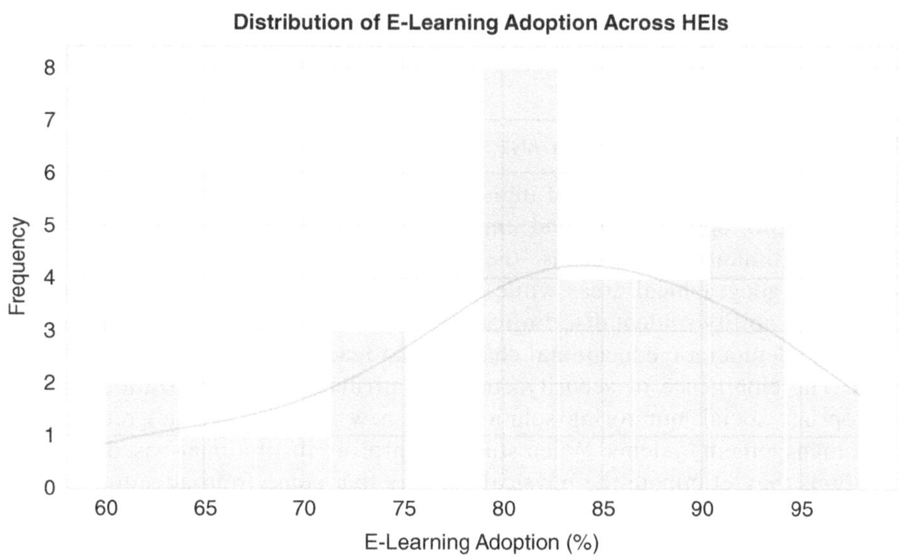

FIGURE 14.5 Distribution of E-learning.

Source: Author's own analysis.

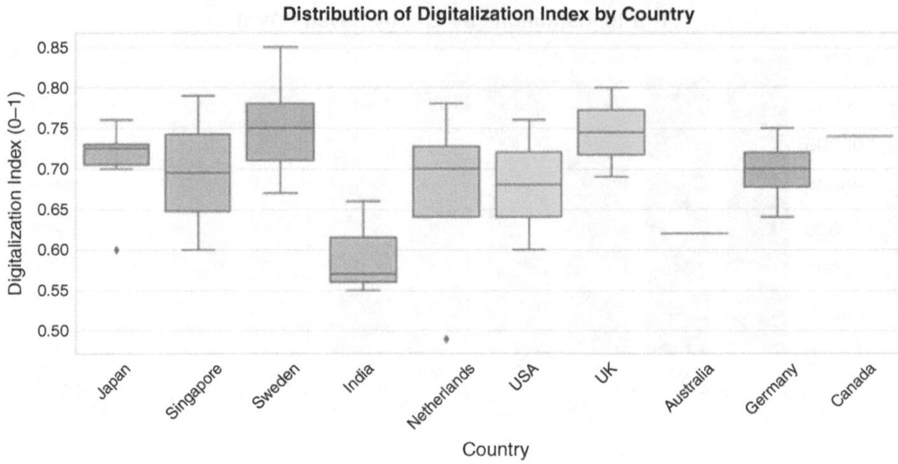

FIGURE 14.6 Distribution of digitalization.

Source: Author's own analysis.

The implementation of digital administration systems enables universities to track grant disbursements and publication indexes to improve transparency by increasing accountability. The improved operational system creates efficiency that decreases academic staff workloads so they can dedicate their time to teaching responsibilities along with scholarship work. Digital solutions enable institutions to enhance financial planning abilities and base all their budgeting processes and procurement activities and alumni fundraising on data. To achieve digitalization through infrastructure development and cybersecurity enhancements and capacity development institutions need to invest initial resources as shown in Figure 14.6 and 14.7.

14.5.3 Social Sustainability and Equity in Access to Education

The main enduring influence of digitalization emerges from the development of equal learning opportunities and improved social sustainability methods. HEIs endure continuous stress because they presently enable learning benefits to persist across geographical areas while keeping social status differences alongside gender and ability student discrimination. Digitalization enables students to gain expanded democratic educational chances that lead to inclusive educational systems. The emergence of security concerns attributed to digitalization requires appropriate social monitoring solutions and new social disparities require efficient management systems. When students enroll in institutional-based inclusion programs they eliminate the physical barriers that come from attending lessons adding endless educational potential. Digital learning platforms enable residents of homes to access online degree programs for their educational completion. The educational formats oriented toward working persons and caregiving families as well as students with disabilities give customized lessons adapted to individual requirements. Educational accessibility improved through digitalization because

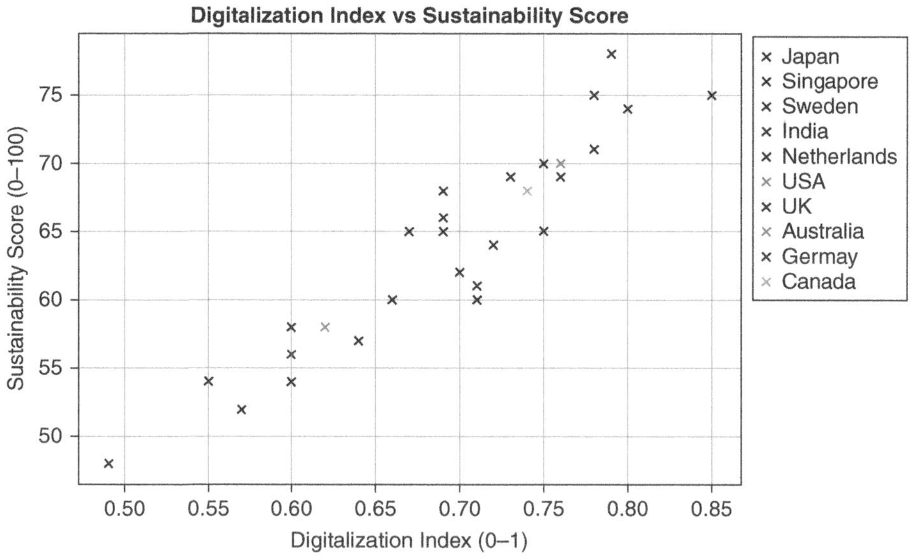

FIGURE 14.7 Digitalization vs sustainability index.

Source: Author's own analysis.

it eliminated obstacles that stood in the way of people seeking higher education enrollment.

Adaptive internet learning platforms integrated with AI educational tools in educational systems allow production of personalized learning content which meets individual requirements. Personalized learning platforms provide assessment metrics which lead students obtaining poor results in traditional education to demonstrate improved results. Online tools provided by students allow them to perform peer reviews which prevent social isolation while facilitating group projects. Digitalization promotes societal sustainability through its capability to facilitate work-based ongoing learning and international friendship development between people across national borders. Students can receive degrees through alternative educational paths because educational programs based on competencies exist online. Social progress occurs because of current workplace movement and persistent educational flexibility needs.

Different social groups experience diverse digitalization effects since they use digital transformation differently. Numerous students cannot benefit from digital learning because digital divisions actively prevent internet access and essential educational equipment together with knowledge-based instructors. Students from deprived economic situations fail to complete their online classes due to their lack of essential digital resources and competencies. Older members of staff struggle to achieve technological adoption success thus reducing the usefulness of educational innovations. Social inequalities find unintended support through the digital approach which leads to the exposure of current inequalities. Toward achieving social sustainability Connecticut State University needs financial support for technology infrastructure as well as educational development and design programs accessibility standards. Universities need to implement active digital tool management systems that produce work efficiency through user-friendly interfaces which enable simple service access for everyone. The process of digitalization creates social equality through knowledge distribution after every societal group gains equal access to these two concepts.

14.6 CASE STUDIES AND BEST PRACTICES

14.6.1 Global Examples of Digital Sustainability in HEIs

Digital sustainability operates as a tested transformative method to help universities globally establish learning centers with complete environmental responsibility systems. Schools beyond their core responsibilities formed efficient programs to create digital sustainability frameworks and platforms. The University of British Columbia (UBC) in Canada builds its entire sustainability strategy through digital system deployment. The UBC executes smart building technology alongside an energy informatics platform for immediate energy assessment across all educational facilities. Predictive maintenance operations involving data processing at the university achieve both cost savings and the highest possible emission reductions. UBC gives internet-based access to its digital sustainability dashboard for public users to provide responses on sustainability measurements. The Netherlands-based Wageningen University achieves its progressive innovations through combining virtual agricultural research facilities and simulations to produce environmental sustainability results without requiring advanced equipment or experimental costs. The university enables its students and researchers to perform simulations that substitute traditional field-based testing through virtual laboratories. The research platforms distribute sustainability science content to students across the globe because of digital resources.

Sustainability objectives traverse the entire Asian campus of the National University of Singapore through its robust digital infrastructure. The facility management aspects at NUS operate through their IoT sensor-based IoT platform that oversees temperature readings and space utilization logs and controls utility usages. The system uses AI-enabled programmed data analysis to stop HVAC system operation when all areas remain unoccupied. Through its digital curriculum program the university establishes sustainability instruction for multiple academic subjects along with digital platforms for university-student collaborative projects. The University of São Paulo (USP) Brazil created an online learning environment which provided extended educational opportunities to reach students located in remote areas and social disadvantage communities. The digital extension programs at USP allowed the university to deliver sustainability content regarding environmental policy together with renewable energy teachings and urban sustainability education to its entire enrollment base of thousands. The digital availability of knowledge allows communities to become powerful sustainability advocates through internet-based solutions which help members solve sustainability challenges they have created. HEIs must recognize digital systems as necessary components of sustainability innovation because they both boost institutional operations and deliver sustainable results across several domains.

14.6.2 Comparative Analysis: Developed vs Developing Countries

Health education institutions follow various digital sustainability approaches because their countries' development depends on infrastructure possibilities and funding accessibility as well as digital platform competency and government backing. Research needs the knowledge gained through these two particular settings to create superior investigations about global digital sustainability. Designed digital systems combined with ample budget support must exist for developed nations to implement full online campus solutions using complex analytics capabilities. Through business partnerships the institution works to develop test and validate upcoming sustainability solutions which combine blockchain technology for carbon tracing and AI solutions to optimize energy use. Different developed nations have implemented emission reduction guidelines that quicken the pace of institutional transformation toward emission reduction objectives. Dominant social institutions choose to direct their funding toward developing sustainable technologies to obtain better hardware development along with enhanced data collection yet remain unconcerned with educational and social system improvements. The current methods used for performance management actively hinder the genuine development of sustainability cultural practices in business organizations.

Academic institutions within developing countries remain unable to apply state-of-the-art technological solutions due to their limitations in digital infrastructure. The educational system limitations pushed scientists to create budget-friendly solutions named frugal innovation which also included adaptive alternatives. The main choice for establishing accessibility among educational institutions in these areas consists of combining open-source tools with mobile-friendly approaches and platforms that scale cost-effectively. Students at African institutions received content messages through SMS from digital learning spaces powered by solar power which operated in off-grid rural areas. The developing region educational institutions create their institutional identity through selecting community service programs from their local areas. Universities implement their teaching methods to develop digital capability bridges with local traditions through instructing environmental conservation techniques to students since digital instruction typically includes comprehensive sustainability training. The interconnected teaching method presents an effective answer to resolve sustainability requirements by establishing local-oriented approaches for water distribution systems and forest conservation as well as urban defense. Developing nations maintain significant sustainable practice capabilities through digital technology despite most advanced infrastructure operating in developed nations. These two essential principles emerge from these paradigms to boost international projects dedicated to capacity development and collaboration objectives.

14.6.3 Lessons Learned and Key Success Factors

Worldwide studies on digitalization and sustainability practices give institutions fundamental concepts to combine various features effectively. An institution reaches its sustainability targets through strategic connections between essential digital transformation goals and particular organizational transformation objectives and academic priorities. Multiple outreach units working together with both academic operations and operational functions create mutual positive effects greater than what any single unit could do alone. Visionary leaders make stability of results achieveable because when they combine their efforts through the team of functional units, this integration generates a clear indication of positive execution. Sustainability officers achieve higher institutional performance when given permission to reach out to IT directors and curriculum designers together with student leaders for developing substantial change programs. The effectiveness of data scientists partnering with digital learning experts and their membership on Python sustainability task forces leads to improved outcomes from solving actual cases (Trevisan et al., 2023).

Organizations developing successful sustainability projects require scheduled evaluation methods for complete organizational transparency. Every area which uses transparency tools and includes digital storytelling characteristics and audit projects needs stakeholder participation for building collective responsibility. Digital sustainability projects gain both institutional capability and deep legitimacy through joint development efforts made by teachers and students. The equity principles in digital sustainability projects need to maintain equal standing with other development direction principles throughout the entire project implementation process. Technology universities need to create distribution systems that provide digital innovation access to individuals with different socioeconomic positions and physical abilities throughout all geographic territories to fulfill social requirements in design phases. Continuous digital sustainability development requires sequential enhancement between teams who team up innovative solutions of different fields.

14.7 FUTURE WORK AND RECOMMENDATIONS

14.7.1 Emerging Technologies in Sustainability: AI, Blockchain, XR, and Beyond

The digital innovation in higher education is in its accelerated stage between testing to whole implementation with digital technologies gradually driving changes in sustainability practices. Artificial intelligence (AI) collaboration with blockchain and extended reality (XR) technologies is an important concept in making innovation more efficient and promising in enhancing sustainable activities in colleges and universities. The technologies are also used to facilitate sustainable operations, research, and pedagogical work in the field of humanities and scientific subjects. Predicting the energy-use patterns in buildings allows promoting the efficiency of operations performed by AI-based analytics and the fact that it is possible to detect the waste of resources in real-time helps minimize the number of natural resources used. Moreover, AI systems can be used to enhance the learning experience by continuously combining the student personality, academic achievements, and local sustainability agendas with personalized course suggestions, such as individual reading resources and learning simulations, which would complement the learning process. Moreover, AI also proves to be of considerable utility in sustainability research by finding use in climate forecasting frameworks, broad ecological data processing, and the creation of environmental policy decision-support systems.

Blockchain technology, though still in its early stages of implementation in HEIs, offers transformative possibilities for sustainability governance. The blockchain system is decentralized, and thus allows institutions to trace carbon footprints while providing approachable origins and metrics based in sustainability for procurement and validation of their metrics. This feature of micro-credentialing on blockchain allows institutions to certify that the achievement of sustainability education does manage to build decentralized lifelong learning programs that are based on the United Nations SDGs. Yet there is growing adoption of educational institutions focused on sustainability, allowing for both virtual and augmented reality system use in the XR. XR technology helps students to learn about the world through virtual scenarios around environmental themes like trees loss, sea levels issues, and the urban development models. It altogether increases learning intensity and mixes emotion. Immersive learning systems form some of the most effective methods of understanding sustainability as complex long-term processes can be converted into the present-oriented form of learning. Falling costs of XR technology mean that the probability of academic adoption for educational purposes and other kinds of needs and programs rises up. The fusion of digital twins and quantum computing and bioinformatics technology creates an opportunity for sustainability research, and digital transformation of institutional strategies. Digital twins allow universities to have its facilities represented in instantaneous digital form to analyze the sustainability improvement methods in simulated environment. Interoperability and increased technology accessibility will result in greater digital sustainability. Institutions are experiencing new ethical and regulatory management difficulties to deal with modern technology systems. Given that new technology is adopted by educational institutions, they ought to set or create rules to safeguard both academic rights and public welfare while they are inclined to environmentally friendly and secure sustainability approaches.

14.7.2 Policy Recommendations for HE Leaders and Governments

The digital innovation progresses from testing through to implementation phases very quickly, digital technologies are currently transforming sustainability practices in higher education. The combination of AI with blockchain and XR technologies plays a very substantial role in making higher education sustainability effective and equitable. It will also have an important role in pursuing sustainable operations as well as humanities and research activities with pedagogical research. The prediction of the energy-use patterns of buildings also allows improving the work of the buildings and decreasing the consumption of natural resources, as AI-based analytics can foresee them and wasteful consumption of the natural resources can also be detected in real-time. AI, making the combination of student characteristics, educational achievements and local sustainability concerns, offers sufficient course recommendations with reading assignments and simulation activities to enrich learning environment. In climate forecasting models, as well as large ecological data assessment, and even in environmental policy creation tools, AI shows how it's worth it in sustainability studies.

Blockchain technology, though still in its early stages of implementation in HEIs, offers transformative possibilities for sustainability governance. The blockchain system is decentralized and thus allows institutions to trace carbon footprints while providing approachable origins and metrics based in sustainability for procurement and validation of their metrics. This feature of micro-credentialing on blockchain allows institutions to certify that the achievement of sustainability education does manage to build decentralized lifelong learning programs that are based on the United Nations SDGs. Yet there is growing adoption of educational institutions focused on sustainability, allowing for both virtual and augmented reality system use in the XR. XR technology helps students to learn about the world through virtual scenarios around environmental themes like trees loss, sea levels issues and the urban development models. It altogether increases learning intensity and mixes emotion. Sequencing of hard to understand processes on the scale of time and scale, such as sustainability, are rendered into present time ones through immersive learning systems which are the most successful approach in sustainability instruction. Falling costs of XR technology mean that the probability of academic adoption for educational purposes and other kinds of needs and programs rises. The fusion of digital twins and quantum computing and bioinformatics technology creates an opportunity for sustainability research, and digital transformation of institutional strategies. Digital twins allow universities to have its facilities represented in instantaneous digital form to analyze the sustainability improvement methods in simulated environment. Interoperability and increased technology accessibility will result in greater digital sustainability. Institutions are experiencing new ethical and regulatory management difficulties to deal with modern technology systems. Given that new technology is adopted by educational institutions, they ought to set or create rules to safeguard both academic rights and public welfare while they are inclined to environmentally friendly and secure sustainability approaches.

14.7.3 Research Gaps and Future Directions

Quick digital transformations are observed in higher education sustainability practices because advanced digital solutions accelerate each step starting from testing to final implementation. Higher education sustainability benefits greatly from the combined power of organic alliances between AI and blockchain technologies and XR platforms. This system plays a critical role in achieving sustainable operations and conducting humanistic and research activities with pedagogical research. AI analytics in building energy forecasting leads to best building operations alongside waste detection results that reduce natural resource utilization. The integration of AI with student features and education results and environmental factors produces suitable course recommendations which contain educational materials and training scenarios to enhance learning quality. Climate forecasting models along with environmental policy development tools and large ecological data assessment make AI an invaluable tool for sustainability research. The increasing number of sustainability-cantered educational institutions creates conditions for using XR systems that contain virtual and augmented reality applications. Students learn about worldwide concepts through XR virtual education which provides interactive simulations about environmental challenges involving tree depletion and rising water elevations and urban planning. The educational intensity of XR increases because emotional connections nurture student learning through advanced technology. Approval-based sustainability instruction implements immersive learning systems because they convert multiple complex processes into contemporary educational content. Twice decreasing costs of XR technology make the adoption of this technology by educational institutions and other organizations now more probable than ever. The integration between digital twins and quantum computing and bioinformatics technology drives both sustainability research and institutional digital transformation. The virtual digital representations of university facilities through digital twins create educational opportunities to assess different sustainability improvements using simulated virtual models. Higher technology accessibility and interoperability together create increased digital sustainability levels. The adoption of modern technology by institutions creates new ethical and regulatory hurdles that need proper management. Academic rights and public welfare protections need to be established in educational settings as new technologies are adopted for environmentally friendly yet secure sustainability purposes.

14.8 CONCLUSION

14.8.1 Summary of Key Insights

The digital tools effectively build sustainable performance in all key areas such as environment, economics, and society for HEIs. The first part of this discussion explained that digitalization means much more than implementing new technologies as it requires institutions to transform digital practices and their objectives. Universities now base sustainability efforts on a system that considers environment protection alongside strong financial condition and equal opportunities for all. Digital technology helps universities lower their energy costs and use less paper while making decisions based on statistical analysis. The system enables organizations to run their daily activities better and create expanded learning platforms plus research environments. Through digital means the education system serves more disadvantaged students while personalizing education and making institutions better match their community's needs.

The examples from many countries reveal how organizations use digital technology to build smart campuses, create open studying platforms, analyze data with AI systems, and spread research across various sites. However different nations develop their strategies the text states that local conditions and fair policies are essential to create equal progress across all nations. The chapter analyzed how AI blockchain and XR will help HEIs become more sustainable in the future. It gave university leaders and decision-makers important suggestions to build a sustainable future by properly aligning their actions with a clear vision. The study recognized where investigations are lacking most and encouraged interdisciplinary teams to study sustainability challenges in universities across multiple countries.

14.8.2 Final Reflections on the Digital-Sustainability Nexus

The digital revolution links perfectly with sustainability and pushes 21st-century universities to face new challenges with enthusiasm. The combined zone needs digital technology combined with environmentally friendly products. Managing institutions must defend their basic values for creating an impact as they face accordance. Universities have to establish fundamental performance standards as they address their responsibilities inside our linked environment. Institutions should adopt technology as an active tool for fulfilling sustainability targets instead of adopting it for simple environmental and ethical initiatives. Universities can easily switch to future data-based strategies while maintaining full commitment to social equality and university standards.

Our future direction depends on effective achievements and work results. Digital transformation will worsen inequality and hurt environmental protection when universities do not equally design their plans and keep measuring performance instead of monitoring vital aspects. Our top duty is to develop systems that make processes run better while maintaining justice principles for both people and our planet. HEIs stand in a singular position to lead digitalization transformation in education. As educational leaders they can demonstrate sustainable digital practices in their teaching activities but also build leaders from their research to make progress in this area. HEIs should use advanced technology to solve problems instead of making technology development their primary goal. The digital and natural systems have linked permanently so we need to redesign our everyday practices as well as education tools. The new approach allows universities to guide and produce transformative results outside of traditional teaching spaces which bring real educational opportunities.

REFERENCES

Akour, M. and Alenezi, M. (2022) "Higher education future in the era of digital transformation," *Education sciences*, 12(11), p. 784. https://doi.org/10.3390/educsci12110784.

Anh, H. *et al.* (2024) "BPR/ISP the effective way for digital transformation in higher education institutions," in *Lecture Notes in Networks and Systems*. Cham: Springer Nature Switzerland, pp. 377–391.

Ben Kasmia, N. and M'hamed, H. (2023) "Digitalization of higher education: Impacts on management practices and institutional development. A literature review," *Conhecimento & diversidade*, 15(39), pp. 56–82. https://doi.org/10.18316/rcd.v15i39.11135.

Gorina, L. *et al* (2023) "Sustainable education and digitalization through the prism of the COVID-19 pandemic," *Sustainability*, 15(8), p. 6846. https://doi.org/10.3390/su15086846.

Leal Filho, W. *et al.* (2024) "Digital transformation and sustainable development in higher education in a post-pandemic world," *International journal of sustainable development and world ecology*, 31(1), pp. 108–123. https://doi.org/10.1080/13504509.2023.2237933.

Mothafar, N.A. *et al.* (2024) "Digital social responsibility towards digital education of international higher education students' institutions: Digital culture as mediator," *Heliyon*, 10(17), p. e36442. https://doi.org/10.1016/j.heliyon.2024.e36442.

Perrin, S. and Wang, L. (2021) "COVID-19 and rapid digitalization of learning and teaching: Quality assurance issues and solutions in a Sino-foreign higher education institution," *Quality Assurance in Education*, 29(4), pp. 463–476. https://doi.org/10.1108/qae-12-2020-0167.

Ramaswamy, M. *et al.* (2021) "Reimagining internationalization in higher education through the United Nations sustainable development goals for the betterment of society," *Journal of studies in international education*, 25(4), pp. 388–406. https://doi.org/10.1177/10283153211031046.

Robertsone, G. and Lapiņa, I. (2023) "Digital transformation as a catalyst for sustainability and open innovation," *Journal of open innovation technology market and complexity*, 9(1), p. 100017. https://doi.org/10.1016/j.joitmc.2023.100017.

Trevisan, L.V. et al. (2023) "Digital transformation towards sustainability in higher education: State-of-the-art and future research insights," *Environment development and sustainability*, 26(2), pp. 1–22. Available at: https://doi.org/10.1007/s10668-022-02874-7.

Umair, S., Waqas, U. and Mrugalska, B. (2024) "Cultivating sustainable environmental performance: The role of green talent management, transformational leadership, and employee engagement with green initiatives," *Work*, 78(4), pp. 1093–1105.

Waqas, U., Umair, S., Mrugalska, B. and Ali Sulaiman, M.A.B. (2025) "Twin transformation: The role of digitalization with artificial intelligence, big data and circular supply chain in circular economy performance for sustainability," *Equilibrium (1689-765X)*, 20(2).

15 Quality Assurance Systems in Higher Education for Sustainability

A Study of Omani Higher Education Institutions

Zaheer Ahmed Khan, Faiza Amir, and Ihsan Ashiq

15.1 THE CONCEPT OF QUALITY ASSURANCE IN HIGHER EDUCATION

Quality assurance (QA) in higher education involves carefully arranged processes and practices intended at safeguarding conformity, credibility, and appropriateness of educational programs. There is an important concept called QA in higher education that involves a systematic, continuous process which aims at evaluating, maintaining, and enhancing education programs and institutions so that they can better fulfill their academic and societal missions. QA is a perpetual process of continuous evaluation and improvement of institutions following set standards and effectively fulfills the educational mission. QA in higher education is a systematic process for maintaining and increasing the quality, credibility, and relevance of educational offerings (Singh et al. 2025). The process requires an organization to evaluate its own learning and teaching activities regularly and make improvements to ensure that institutions meet the requirements set by institutional standards and operate effectively (Makhoul 2019). The ultimate objective of policies of an educational institution is to improve achieving learning outcomes and learning experiences of students. This objective serves as the foundation for policies and procedures that promote high-quality instruction (Hénard and Roseveare 2012, 9). There are different standards worldwide, including global, regional, and national. Institutions set their border standards. Globally, economically developed nations have developed the quality and credibility of their tertiary institutions in line with global trends to promote the transformable role of education authorities (UNESCO 2025). In these countries higher education institutions (HEIs) follow institutional standards and continually improve their educational offerings. Global and regional bodies work as advisories in establishing quality systems. The European Higher Education sector employs standards and guidelines for

DOI: 10.1201/9781003611905-21

QA system for transparency and mutual recognition of qualifications (ENQA 2015). Typically, an institution's internal QA consists of governance, curriculum assessment, faculty advancement, and student feedback examinations to maintain the effectiveness (Altbach 2016). These frameworks, together with practice, support the accountability principle, students' learning outcomes, and public trust. Global and regional quality standards facilitate educational benchmarking, ensure institutional credibility, and international comparisons in higher education. QA in higher education maintains and enhances quality through credible and relevant academic offering and practices as a systematic process. Therefore, the effectiveness of QA practices must ensure that HEIs meet established standards consistent implementation, continuous improvement, and accountability.

15.2 QUALITY ASSURANCE FRAMEWORKS AND STANDARDS

QA frameworks and standards are structured approaches for QA to assess and improve the quality of HEIs. QA enables HEIs to evaluate and enhance the quality of educational excellence, continuous improvement, and accountability. These frameworks operate at local, national, and international levels. HEIs' internal QA mechanisms include curriculum reviews, faculty evaluations, and student feedback. HEIs align their practices with accreditation requirements, creating a positive outcome and reputation (Tariq et al. 2025). Global QA frameworks, such as the European Association for Quality Assurance in Higher Education and the International Network for Quality Assurance Agencies in Higher Education (INQAAHE) facilitate cross-border recognition of qualifications. These standards promote consistency and credibility in higher education worldwide (Jafarov 2024).

QA frameworks enhance transparency, student learning outcomes, and institutional competitiveness, ensuring that HEIs meet the evolving educational demands. Standards and frameworks for QA in higher education ensure academic quality, accountability, and continuous improvement. Globally, QA frameworks operate at international, regional, and national levels, fostering consistency and credibility in higher education systems.

15.3 INTERNATIONAL FRAMEWORKS

Organizations, such as UNESCO and the OECD, have developed guidelines to promote consistent quality in higher education across borders. The guidelines on quality provisions at the international level aim to alleviate low-quality education through international cooperation among governments, institutions, and QA agencies by addressing the responsibilities of various stakeholders in ensuring quality education (Vincent-Lancrin and Pfotenhauer 2012). International frameworks such as the Bologna Process, the Lisbon Recognition Convention, and UNESCO Global Convention on Higher Education contribute greatly to the standardization and enhancement of higher education worldwide. By developing a coherent and compatible European Higher Education Area (Bouhlal et al. 2019), the Bologna Process aims to improve student mobility and employability. Since 1999, the Lisbon Recognition Convention has ensured mutual recognition of qualifications throughout Europe and beyond, as well as QA. A global convention on higher education adopted in 2019 by UNESCO fosters international cooperation and trust among institutions and recognizes higher education qualifications globally (Teter et al. 2021).

15.4 REGIONAL FRAMEWORKS

In higher education, regional frameworks establish accreditation standards, foster academic collaboration, and promote mutual recognition of qualifications. These frameworks facilitate benchmarking, enhance curriculum alignment, and support mobility programs. With the help of frameworks such as the Bologna Process (Europe) and AQAF (ASEAN), education systems can increase institutional accountability and continuously improve their quality. The EHEA implements standards and guidelines for QA at the regional level. It promotes transparency and mutual recognition across Europe. The ASEAN Quality Assurance Framework works for regional collaboration to harmonize educational standards in Asia (Bateman 2016). Regional frameworks work in specific contexts to address local educational needs and a regulatory environment. The EHEA, for instance, provides guidelines and standards for QA within the European Higher Education Area as a framework for internal and external QA. Similarly, Brisbane Communiqué supports collaboration among countries in the Asia-Pacific region. Developing QA frameworks and linking them to international standards will facilitate students' mobility.

15.5 COUNTRY LEVEL FRAMEWORKS

Several regulatory bodies operate at the national level, which act as QA standards for higher education. For example, in the United Kingdom, the Quality Assurance Agency, the Oman Authority for Academic Accreditation and Quality Assurance (OAAAQA) in Oman, and the National Assessment and Accreditation Council (NAAC) in India certify HEIs. In Malaysia, the Malaysian Qualifications Agency (MQA) certifies the institutions and degree programs. These bodies ensure that HEIs in the country meet predefined standards and promote consistency, transparency, and accountability (Naim et al. 2024).

To maintain and enhance the standards of HEIs, country-level QA frameworks play an important role. They establish national benchmarks for curriculum development, faculty qualifications, research output, and institutional governance to ensure consistency and credibility in education. These frameworks have accreditation mechanisms, hold institutions accountable, are transparent, and incessantly improve their academic and professional experience (Altbach et al. 2019). One of the emphases of QA frameworks is to facilitate the development of skills relevant to industry and society and align education with national development goals (Martin and Stella 2018). Additionally, they support degree recognition internationally, which assists the relocation of students anywhere in the world. QA frameworks prevent substandard institutions from misleading practices of substandard institutions. QA measures by local QA frameworks enhance higher education's legitimacy (Hsu et al. 2024).

15.6 SUSTAINABLE DEVELOPMENT IN HIGHER EDUCATION

Sustainability in higher education integrates sustainable practices and values into the educational process in universities and colleges. Sustainability aims to address present needs without compromising the ability of future generations to meet them. This concept aligns with economic growth, social inclusion, and long-term resource protection (Abo-Khalil 2024; UNESCO 2020). Therefore, in higher education, sustainability incorporates principles in teaching, research, and operational performance through community engagement. It prepares students to understand complex global processes and promotes innovation and responsible citizenship for the prosperity of society and generations ahead. Sustainability in higher education goes beyond economic viability, social equity, and cultural inclusivity. Sustainability in higher education includes embedding sustainability objectives into courses, campus operations through sustainable infrastructure and use of resources, community engagement, and governance. Another key dimension of sustainable practices is research and innovation at HEIs (Iqbal and Piwowar-Sulej 2022). Education for Sustainable Development (ESD) provides students with the knowledge, skills, and attitudes required to contribute to a sustainable society. ESD encourages critical thinking, problem-solving, and interdisciplinary learning, enabling students to comprehend and confront interrelated global issues. It promotes a transition from traditional, discipline-specific instruction to a more integrated approach that emphasize systems thinking and real-world applications (Edwards et al. 2020). By promoting responsible citizenship, ESD empowers students to make informed decisions and engage in responsible actions that support sustainable development.

15.7 HIGHER EDUCATION IN ACHIEVING SUSTAINABLE DEVELOPMENT GOALS (SDGs)

HEIs act as drivers of global sustainable development by promoting literacy, research, and modeling sustainable practices (Elmassah, Biltagy and Gamal 2022, 330). HEIs contribute to the following SDGs: (1) Quality Education (SDG 4), (2) Climate Action (SDG 13), and (3) Partnerships for the Goals (SDG 17). HEIs embed sustainable development goals through scholarly activities and their participation of HEIs makes them attainable (Adhikari and Shrestha 2023; Chankseliani and McCowan 2021, 1115) and social impact through the following initiatives.

15.7.1 Educating Future Leaders

Academic programs develop attributes to prepare future leaders by teaching sustainability principles in relevant courses from various disciplines. This initiative develops competence, knowledge, and ethical responsibility among graduates to motivate them. Sustainability education promotes thoughtful inquiry, problem-solving, and systems-based approaches to challenging global issues and engage students into change as initiative-taking agents.

15.7.2 Conducting Research

Academic research is a building block of sustainable development. The scientific research generates knowledge, technological advancements, innovation, and provides policy recommendations. HEIs play a significant role in tackling challenges such as climate change, poverty alleviation, clean energy, infrastructure development, and social justice. The research by HEIs forms evidence-based solutions to feed national and international sustainability policy frameworks.

15.7.3 Strategic Partnerships

The partnership between HEIs, local industries, nonprofit organizations, and international organizations support advancement in sustainability. Such partnerships have local, national, and global impacts. Strategic partnerships facilitate knowledge flows, expertise, and resources, and promote sustainability interventions. Joint efforts by HEIs and other actors encourage engagement of community, governments, and public through research to real-life outcomes.

15.7.4 Sustainable Practices

HEIs demonstrate sustainable practices through various initiatives to minimize the environmental impact of their campus operations. HEIs integrate sustainability-focused curricula and research, foster innovation, promote energy conservation, and promote sustainable agriculture. There are also student-led initiatives, such as eco-friendly transportation and community outreach programs to promote environmental awareness. HEIs opt for ethical investments to support socially responsible ventures of eco-conscious projects, sustainable domestic agriculture, and energy conservation. Other important outcomes of HEI research include reducing carbon footprints, disposing of waste, building green infrastructure, using renewable energy, and developing responsible procurement. These efforts enable HEIs to play a significant role in exemplifying sustainable practices to future generations and the broader community.

15.8 QUALITY ASSURANCE PRACTICES AND CHALLENGES

QA in HEIs has become a global priority, focusing on successful practices worldwide, addressing usual challenges, and implementing strategic solutions. QA integrates collaborations to embrace internationalization and technology, ensuring the creation of balanced regulations that lead to excellence in higher education. Global accreditation bodies implement latest approaches in accreditations as an exemplary practice, while local and regional QA frameworks align their standards with international benchmarks. Typically, global best practices emphasize continuous improvement and long-term planning, monitored through key performance indicators (KPIs). HEIs develop robust data management systems to support informed decision-making and maintain quality standards (Javed and Alenezi 2023). Despite significant achievements, institutions still face the following challenges in implementing effective QA systems (Adams 2025; Gulua 2020; QAA 2022).

15.8.1 Resource Constraints

In many cases, limited resources hinder HEIs' ability to maintain robust QA practices. These constraints are typically financial, human resources, or capacity. HEIs usually lack sufficient infrastructure, updated technology, and qualified personnel to maintain QA at a desirable standard.

15.8.2 Regulatory Burdens

The complexity of regulatory requirements can sometimes challenge HEIs, especially small colleges. For example, conflicting or overlapping requirements and globalization versus local standards can be annoying. Such complications increase significant financial and administrative strains on HEIs.

15.8.3 Balancing Innovation and Compliance

The alignment of innovative teaching methods with existing QA frameworks may pose challenges for institutions seeking to pursue innovative teaching and learning methods. Implementing new educational practices can be challenging because of the tension between fostering innovation and adhering to established quality standards.

15.9 STRATEGIES TO OVERCOME CHALLENGES

15.9.1 Leveraging Technology

Implementing advanced technology solutions can streamline QA processes and reduce administrative burdens. For instance, developing comprehensive data management systems enables efficient monitoring of KPIs and supports evidence-based decision-making (Javed and Alenezi 2023).

15.9.2 Collaborative Approaches

Engaging in partnerships with other institutions, QA agencies, and international bodies can provide access to shared resources, expertise, and best practices. Collaborative efforts can enhance the capacity of institutions to meet QA standards and foster a culture of continuous improvement (Al-Khouri 2023).

15.9.3 Regulatory Reform Advocacy

Institutions can collectively advocate for regulatory frameworks that balance accountability with flexibility. By voicing concerns and proposing solutions, higher education leaders can influence policies that support effective QA without imposing undue burdens (Adams 2025).

15.9.4 Integrating Innovation into QA Frameworks

Updating QA processes to accommodate and encourage innovative practices ensures that QA evolves alongside scientific, technological, and educational advancements. This integration is achievable by developing guidelines that assess the quality of innovative teaching methods and curricular designs (QAA 2022).

15.10 QUALITY ASSURANCE IN HIGHER EDUCATION IN OMAN

The need for global alignment, international competitiveness, and educational quality motivates QA at HEIs in Oman to improve student learning and institutional performance. Oman's higher education system has grown and evolved since the 1970s from a small number of options to a diversified industry, including both public and private universities and colleges. The regulatory frameworks and accrediting procedures support preserving quality, faculty development, and institutional effectiveness (Alyahmadi 2007; Al Amri et al. 2020). Various bodies regulate QA in Oman. The Ministry of Higher Education, Research, and Innovation (MoHERI) has a significant role in shaping the strategic direction for QA in higher education. Educational policies and guidelines come from MOHERI to build a culture of continuous improvement and academic excellence in HEIs. The MoHERI collaborates with the OAAAQA, being the primary body in higher education and other stakeholders, to maintain ongoing evaluations to enhance the quality of higher education in Oman. OAAAQA assures quality standards through evaluations and reviews for institutional accreditation and program accreditation (Oman Academic Accreditation Authority [OAAA] 2021). OAAAQA awards institutional accreditation, and the Oman Qualifications Framework (OQF) awards program accreditation. In the last two decades, higher education in Oman has undergone several development phases. Qualifications from both the public and private sectors of education and training in Oman include in OQF mandate, integrated and required National Qualifications Framework (NQF). HEIs in the Gulf region are also adopting international QA standards, often in compliance with US, UK, or EU accrediting agencies (Shahzad and Shaukat 2021). Some Omani HEIs seek international accreditation from the Accreditation Board for Engineering and Technology (ABET) or the Association to Advance Collegiate Schools of Business (AACSB).

15.11 QUALITY ASSURANCE FRAMEWORK IN OMAN

Component	Description	Source
Regulatory Body	The Royal Decree 9/2021 created Oman Authority for Academic Accreditation and Quality Assurance of Education (OAAAQA). The body is responsible to oversee the accreditation and quality assurance of HEIs in Oman.	(Oman Authority for Academic Accreditation and Quality Assurance of Education 2024)
Institutional Accreditation	Institutional Accreditation is an outcome; ISA is the process used by the OAAAQA to determine if an HEI has achieved this outcome. It is an independent evaluation of the extent to which an HEI meets the national Standards expected of all HEIs in Oman and to confirm the good standing of an institution. The ISA consists of a self-assessment by the HEI and an external assessment by an expert Panel appointed by the OAAAQA.	
Institutional Standards	1. Governance and Management 2. Academic Provision and Resources 3. Student Support Services 4. Human Resources 5. Research and Innovation 6. Partnerships and External Engagement	
Assessment Methodology	ADRI approach (Approach, Deployment, Results, and Improvement) institutional self-assessment report. Multiple evidence sources (documentation), interviews, and site visits	
Institutional Accreditation Outcomes	Institutions' accreditation categories: 1. Accredited (all standards met) 2. Conditionally Accredited (five out of six standards met) 3. On Probation (fewer than five standards met)	
Review and Appeals Process	Institutions may submit an appeal for review on accreditation outcomes. OAAAQA oversees appeals according to national laws and international quality assurance standards.	
External Quality Assurance (EQA) Activities	• Institutional Standards Assessment (ISA) • Program Standards Assessment (PSA) • Institutional and Program Reassessment (ISR and PSR) • General Foundation Program Quality Audit (GFPQA) • Reviews of International EQA (IEQA) applications	
ISA Process and Timeline	• Timeline for Institutional Standards Assessment (ISA) • Self-assessment submission, external review, site visits • Final accreditation decision • Approximate time 28 weeks	
Oman Qualifications Framework (OQF) Qualification System	OQF defines the following: • Qualification levels, credit hours, learning outcomes, assessment criteria • Pathways for academic, professional, technological, and vocational education • Credit transfer and Recognition of Prior Learning (RPL) • Program listing • Program standard accreditation	(OAAAQA 2023, 7, 12–16)

15.12 IMPACT OF QUALITY ASSURANCE PRACTICES IN HEIs IN OMAN: EVIDENCE FROM LITERATURE

The higher education sector in Oman is rapidly growing and embracing the best practices for quality enhancement. Higher education in Oman has integrated QA mechanisms to maintain a high standard of education and keep the higher education system aligned with global benchmarks. This literature analysis explores the impact of QA in higher education in Oman and presents a thematic analysis of relevant studies of the past five years (2018–2023). After reviewing and examining the selected publications related to the impact of QA on HEIs in Oman, themes uncover the key findings related to QA in Oman's higher education. The thematic analysis revealed key findings, including improvement of educational standards. The study examined the influence accreditation has on institutional practices, the challenges associated with implementation, the positive impact on teaching and learning, and the improved global competitiveness of Omani HEIs.

15.13 METHODOLOGY

To assess the outcomes of QA practices in HEIs in Oman, this study conducted a thematic search using academic databases such as Google Scholar, Scopus, and JSTOR, considering publications of five years (2018–2023). The study used the following keywords for search: "quality assurance in higher education Oman," "impact of QA in Omani universities," "Oman academic accreditation," and "higher education reforms Oman." The study included articles that directly examined the impact of QA on educational practices, institutional performance, or student outcomes concerning HEIs in Oman. We excluded articles that did not focus on Oman or QA in HEIs. The study analyzed the selected studies through thematic analysis to identify recurring insights and themes.

15.14 THEMATIC ANALYSIS

The selected studies identified the following recurring themes related to the impact of QA in higher education in Oman.

Theme	Description	Citations
1. Improvement of Educational Standards	By implementing quality assurance mechanisms, the Oman Academic Accreditation Authority (OAAA) has significantly contributed to improving academic rigor, curricula, and policies. At the institutional level, the adoption of international best practices is increasing to align with global education standards.	(Al-Dahash and Al-Busaidi 2020; Al-Shihi and Al-Farsi 2019; Al Shanfari 2017)
2. Institutional and Programmatic Accreditation	Resulting of the accreditation drive in Oman, institutions are adopting self-assessment mechanisms, improve faculty development and student engagement. HEIs also seek international accreditation to ensure compliance with global standards.	(Al-Busaidi and Al-Harthy 2020; Al-Shihi and Al-Farsi 2019; OAAAQA 2021)

(Continued)

Theme	Description	Citations
3. Challenges in Implementing Quality Assurance	Some institutions face resource limitations, faculty resistance, and a lack of quality assurance cultures. A lack of funding, inadequate faculty training, and misalignment between QA policies and HEIs operations contribute to implementation difficulties.	(Al-Busaidi and Al-Harthy 2020; Al-Dahash and Al-Busaidi 2020)
4. Impact on Teaching and Learning Practices	The implementation of QA mechanisms has improved teaching quality, faculty development, adoption of modern teaching methods, and student-centered approaches. Institutions use student feedback actively to improve the quality of instruction.	(Al-Dahash and Al-Busaidi 2020; Al-Shihi and Al-Farsi 2019)
5. Global Competitiveness and International Recognition	By partnering internationally, exchanging students, and collaborating in research, Omani HEIs have enhanced their global reputation. Higher education institutions are becoming more competitive in the global education market through international quality standards.	(Al-Busaidi and Al-Harthy 2020; Al-Saadi 2023; UNESCO & LRCC 2010)

15.15 DISCUSSION

Evidence from the literature demonstrates that QA systems have helped to improve the standards of higher education in Oman. A structured approach established by the OAAA has helped evaluate and improve academic quality. QA frameworks promoted academic rigor, curricula, and institutional effectiveness. QA initiatives have helped streamline processes and ensure compliance with national and international benchmarks. Moreover, it has promoted evidence-based decision-making, which helped to improve student learning outcomes and an effective academic environment. OAAAQA has accredited several institutions and programs that have significantly impacted the quality standards and programs. QA mechanisms have been helpful in institutional evaluation, program evaluation, faculty development, and student engagement.

Participation of faculty members in QA has enhanced their professional development, the use of modern instructional methods, and the use of valid assessment tools aligned with global standards. This participation has resulted in enhanced classroom experience and student satisfaction. Also, QA practices have enhanced opportunities for collaborations with reputed global institutions. Hence, by maintaining rigorous educational standards, Omani HEIs can compete in the outer world higher education marketplace to attract domestic and international students.

Despite notable successes, Omani institutions are experiencing certain challenges in implementing QA practices. These challenges are (1) resource limitations, (2) resistance to change, and (3) an underdeveloped QA culture. Furthermore, some institutions have had difficulty adopting new QA frameworks due to insufficient funding, insufficient faculty training, and misalignment between daily operations. Establishing a culture change toward the importance of QA in education is one difficulty with some HEIs. Although Omani HEIs have achieved performance indicators of QA locally, some employees and academicians still view it as a formality rather than a chance to improve.

15.16 FUTURE PERSPECTIVES ON QUALITY ASSURANCE IN OMANI HIGHER EDUCATION

Due to the continued prioritization of educational reforms, the future of QA effectiveness in Omani HEIS looks promising. The integration of digital tools in teaching and faculty participation in QA training with OAAAQA is enhancing the effectiveness of institutions. Both MOHERI and OAAAQA have increased collaboration with international accrediting bodies that will raise Oman's competitiveness in the global education landscape. QA processes in higher education in Oman expect transformation by embracing practices of international collaborations, digital integrations, and regulatory refinements. There is steady progress among HEIs in Oman in securing OAAAQA accreditation and program listing for program accreditation in the future. The rise in efforts for dual accreditation by HEIs in one (locally and globally) indicates an elevation of institutional standards. The integration of virtual collaboration tools will become more prevalent to enhance QA (Al-Dahash and Al-Busaidi 2020). Furthermore, continuous refinement of QA processes by authorities in Oman is moving toward more dynamic, real-time accreditation methods rather than periodic reviews. Omani HEIs are focusing on research-driven education that can contribute to local and global knowledge, enhance their academic reputation, and align with international quality benchmarks. Omani HEIs are endeavoring to align with international QA standards through diverse strategies: Future strategies may involve (1) partnering with global QA agencies and cross-border QA networks. (2) Facilitating students and faculty through exchange programs. Expectedly, this will open new venues to diversify learning experiences for students and HEIs.

15.17 CONCLUSION

QA practices and mechanisms in higher education in Oman rely on the philosophical approach of QA in higher education. QA mechanisms and practices in Oman aim to help local HEIs become competitive in the global higher education landscape. Also, QA initiatives in higher education in Oman stem from the long-term goals of Vision 2040 of Oman as a national strategy to develop a broad base and diverse economy. QA is one of the pillars of higher education in Oman and endeavors to safeguard the quality and relevance of the institutions. By maintaining QA standards, HEIs in Oman can compete globally.

QA processes have remarkably achieved quality standards in Omani HEIs through their regulatory frameworks. Still, some challenges remain, including inadequate resources, a culture of resistance to changes, and creating a balance between international and local standards. The transition of using technology with research-driven education would further enhance the evolution of the higher education sector in Oman.

REFERENCES

Abo-Khalil, A. G. (2024). Integrating sustainability into higher education: Challenges and opportunities for universities worldwide. *Heliyon*, *10*(9).

Adams, R. (2025, February 16). University leaders in England call for rethink over rising 'regulatory burden'. *The Guardian*. https://www.theguardian.com/education/2025/feb/16/university-leaders-england-regulator-fees-office-for-students

Adhikari, D. R., & Shrestha, P. (2023). Knowledge management initiatives for achieving Sustainable Development Goal 4.7: Higher education institutions' stakeholder perspectives. *Journal of Knowledge Management*, *27*(4), 1109–1139.

Al-Amri, A. S., Mathew, P., Zubairi, Y. Z., & Jani, R. (2020). Optimal standards to measure the quality of higher education institutions in Oman: Stakeholders' perception. *SAGE Open*, *10*(3), 2158244020947440.

Al-Busaidi, K. A., & Al-Harthy, A. M. (2020). Challenges in the implementation of quality assurance in Omani higher education institutions. *International Journal of Educational Development*, *70*, 102167. https://doi.org/10.1016/j.ijedudev.2020.102167

Al-Dahash, H. A., & Al-Busaidi, K. A. (2020). The impact of digital learning on quality assurance in higher education: A case study of Omani institutions. *Journal of Educational Technology & Society*, *23*(4), 126–137.

Al-Khouri, A. M. (2023). A case study on institutional best practices for quality assurance in a research university. *International Journal of Educational Development*, *92*, 102115.

Al-Saadi, Z. T. (2023). A review of graduate attributes in the Oman Authority for Academic Accreditation and Quality Assurance of Education (OAAAQAE) quality audit reports. *Gulf Education and Social Policy Review*, 107–124.

Al-Shanfari, S. A. (2017). *Transnational higher education and quality: Oman's experience and the concept of policy borrowing*.

Al-Shihi, H., & Al-Farsi, M. (2019). International accreditation in Omani higher education institutions: A challenge or an opportunity? *Higher Education Policy*, *32*(3), 375–392.

Altbach, P. G. (2016). *Global Perspectives on Higher Education*. Johns Hopkins University Press.

Altbach, P. G., Reisberg, L., & Rumbley, L. E. (2019). *Trends in Global Higher Education: Tracking an Academic Revolution* (Vol. 22). Brill.

Alyahmadi, H. H. (2007). *The development and expansion of higher education in Oman: Planning for the future* (Doctoral dissertation, University of Sheffield).

Bateman, A. (2016). *ASEAN guiding principles for quality assurance and recognition of competency certification systems*.

Bouhlal, A., & Wells, P. J. (2019). UNESCO conventions on the recognition of higher-education qualifications. In *Major Challenges Facing Higher Education in the Arab World: Quality Assurance and Relevance* (pp. 225–236). Springer International Publishing.

Chankseliani, M., & McCowan, T. (2021). Higher education and the Sustainable Development Goals. *Higher Education*, *81*(1), 1–8.

Edwards Jr., D. B., Sustarsic, M., Chiba, M., McCormick, M., Goo, M., & Perriton, S. (2020). Achieving and monitoring education for sustainable development and global citizenship: A systematic review of the literature. *Sustainability*, *12*(4), 1383.

Elmassah, S., Biltagy, M., & Gamal, D. (2022). Framing the role of higher education in sustainable development: A case study analysis. *International Journal of Sustainability in Higher Education*, *23*(2), 320–355.

European Association for Quality Assurance in Higher Education. (2015). *Standards and guidelines for quality assurance in the European Higher Education Area (ESG)*. ENQA. https://enqa.eu/index.php/home/esg/

Gulua, E. (2020). Barriers to quality assurance for higher education institutions. *Journal of Education and Practice, 11*(12), 45–56. https://doi.org/10.7176/JEP/11-12-05

Henard, F., & Roseveare, D. (2012). *Fostering quality teaching in higher education: Policies and practices.* OECD. https://www.oecd.org/education/imhe/QT%20policies%20and%20practices.pdf

Hsu, Y.-P., & Xu, R.-H. (2024). Problematising the quality assurance policy shift in Taiwan: From external towards internal accreditation mechanisms. *Higher Education Policy, 37*(1), 191–207.

Iqbal, Q., & Piwowar-Sulej, K. (2022). Sustainable leadership in higher education institutions: Social innovation as a mechanism. *International Journal of Sustainability in Higher Education, 23*(8), 1–20.

Jafarov, S. (2024). Accreditation and quality assurance in world-leading universities. *International Journal of Current Science Research and Review, 7,* 12.

Javed, Y., & Alenezi, M. R. (2023). A case study on sustainable quality assurance in higher education. *Sustainability, 15*(10), 8136. https://doi.org/10.3390/su15108136

Martin, M. (2018). *Internal Quality Assurance: Enhancing Higher Education Quality and Graduate Employability.* UNESCO.

Miller, M. T. (2016). *Improving Teaching and Learning: An Overview of the Process.* Education Press.

Naim, A., Saklani, A., Khan, S. A., & Malik, P. K. (Eds.). (2024). *Evaluating Global Accreditation Standards for Higher Education.* IGI Global.

Oman Academic Accreditation Authority. (2021). *Annual report 2020.* Muscat: Author.

Oman Authority for Academic Accreditation and Quality Assurance of Education. (2023). *Oman Qualifications Framework manual (OQF Manual).* https://nrq.oaaaqa.gov.om:20443/assets/documents/oqf_manual.pdf

Oman Authority for Academic Accreditation and Quality Assurance of Education. (2024). *Institutional standards assessment manual (ISAM v2).* https://www.utas.edu.om/portals/2/booklet/02.%20Institutional%20Standards%20Assessment%20ISAMv2.pdf

Quality Assurance Agency. (n.d.). *When quality assurance meets innovation in higher education.* https://www.qaa.ac.uk/docs/qaa/members/when-quality-assurance-meets-innovation-in-higher-education.pdf

Shahzad, K., & Shaukat, M. R. (2021). Quality prospects in higher education: A case of Oman. In *Higher Education in the Gulf* (pp. 109–122). Routledge.

Singh, J. P., Mishra, N., & Mishra, K. K. (2025). *Adapting Quality Assurance Practices in Higher Education for the 21st Century: Challenges and Innovations.* IGI Global Scientific Publishing.

Tariq, M. U. (2025). Sustainability of quality processes in higher education: Strategies for continuous improvement. In *Higher Education and Quality Assurance Practices* (pp. 305–334). IGI Global Scientific Publishing.

Teter, W. R., & Wang, L. (2021). Monitoring implementation of the Tokyo Convention on recognition: A multi-stakeholder approach to the internationalization of higher education in the Asia-Pacific. *International Journal of Comparative Education and Development, 23*(3), 157–174.

UNESCO. (2010). *Revised recommendation on the recognition of foreign qualifications.*

UNESCO. (2020). *Education for sustainable development: A roadmap.* https://unesdoc.unesco.org/ark:/48223/pf0000374802

UNESCO. (n.d.). *Guidelines on quality provision in cross-border higher education.* https://unesdoc.unesco.org

Vincent-Lancrin, S., & Pfotenhauer, S. (2012). *Guidelines for quality provision in cross-border higher education: Where do we strand?* OECD Education Working Papers No. 70. OECD Publishing.

Index

D

E